Lecture Notes in Mathematics

A collection of informal reports and seminars
Edited by A. Dold, Heidelberg and B. Eckmann, Zürich

231

Hans Reiter

Mathematisch Instituut der Rijksuniversiteit
De Uithof, Utrecht/Nederland

L^1-Algebras and Segal Algebras

Springer-Verlag
Berlin · Heidelberg · New York 1971

AMS Subject Classifications (1970): 43 A 20, 43 A 25, 43 A 45

ISBN 3-540-05651-3 Springer-Verlag Berlin · Heidelberg · New York
ISBN 0-387-05651-3 Springer-Verlag New York · Heidelberg · Berlin

Offsetdruck: Julius Beltz, Hemsbach/Bergstr.

PREFACE

L^1-algebras of locally compact groups have been studied for many years. The systematic study of Segal algebras - which are a generalization of L^1-algebras - was begun only a few years ago, for locally compact abelian groups. It is the purpose of these notes to study Segal algebras for general locally compact groups and to discuss also some new results for L^1-algebras. There are many problems here for further research, and it is hoped that the reader will be able to continue where the author had to stop.

The lectures of which these notes are the outcome were delivered at the Universities of Heidelberg, Nancy and Utrecht in 1969 and 1970. I wish to express here my cordial thanks to Professors H. Leptin (Heidelberg) and P. Eymard (Nancy) for their invitations and, especially, to my friends and colleagues in Utrecht who have made my long stay in the Netherlands so pleasant.

Thanks are also due to Professor B. Eckmann of the Eidgenössische Technische Hochschule, Zürich, and to Springer-Verlag for making publication of these lectures possible.

Dr. W. Beiglböck (Heidelberg), Mr. J.-P. Pier (Nancy) and Mr. M. Riemersma (Utrecht) have kindly provided me with their notes of the original lectures; Mr. Riemersma has also examined the manuscript itself and has contributed much helpful criticism. Last but not least, Miss W. Jenner (Utrecht) has typed the whole manuscript with the utmost care. My thanks go to each and all of them.

July 1971 H.R.

Mathematisch Instituut der Rijksuniversiteit
De Uithof, Budapestlaan
UTRECHT, The Netherlands

Address from 1 September 1971:
Mathematisches Institut der Universität,
Strudlhofgasse 4
A - 1090 VIENNA, Austria

READER'S GUIDE

The reader is assumed to be acquainted with the author's monograph '<u>Classical harmonic analysis and locally compact groups</u>' (Oxford University Press, 1968); terminology, notation and results explained there are used freely here.

<u>References</u> such as Ch. 6, § 2.2, without further specification, refer to the monograph mentioned. References to the <u>Bibliography</u> at the end of the present lecture notes are given by a name in capital letters followed by a number in brackets, e.g. WEIL [1].

The <u>subject-matter of the lectures</u> may be divided into three parts (which are closely interrelated):

(i) A study of certain ideals of $L^1(G)$ associated with closed subgroups of G. Here the property P_1 is shown to play an essential role. This part represents a generalization of results given in REITER [2].

(ii) The investigation of Segal algebras $S^1(G)$ for general locally compact groups. This extends the study of the abelian case in Ch. 6, § 2. Ideals associated with subgroups of G are studied for Segal algebras.

(iii) A general approach to approximate units, especially in closed ideals of $L^1(G)$ or $S^1(G)$. This is connected with recent work in harmonic analysis described in Ch. 7, § 4, and leads to new developments.

A <u>summary of contents</u> follows.

In § 1 we associate with a closed subgroup H of a locally compact group G a closed left ideal and a closed right ideal of $L^1(G)$. These ideals coincide if and only if H is normal; they play a central role in later developments.

§ 2 contains some preliminary applications of the property P_1 to the ideals introduced in § 1.

In § 3 we discuss a modification of the property P_1, and also its relativization (in the sense of Ch. 4, § 5.1). This is needed for later applications.

In § 4 Segal algebras are defined for general locally compact groups; they may be said to constitute a natural extension of L^1-algebras There are two main classes: symmetric and pseudosymmetric Segal algebras the former are closer to the abelian case. Some basic properties of Segal algebras are discussed; here integration of functions with values in a Banach space plays a role.

In § 5 examples of Segal algebras are given; some of these present non-trivial problems (cf. examples (iii) and (viii) in § 5).

In § 6 approximate left (right, two-sided) units are introduced, for Banach algebras. This concept, as defined here, is more general than that traditionally used; the significance of the general definition will emerge from the results to be proved in these lectures.

§ 7 contains some lemmas on approximate units of various kinds, in Banach algebras. These lemmas are required for later applications; the proofs are somewhat technical and may well be omitted on a first reading.

In § 8 approximate units in Segal algebras are considered. It is shown that some familiar properties of $L^1(G)$ carry over to Segal algebras, but the proofs are more subtle than in the classical case of $L^1(G)$.

§ 9 is devoted to a general study of closed ideals in Segal algebras. It is proved that the bijective correspondence between the closed ideals of a Segal algebra $S^1(G)$ and those of $L^1(G)$, established in Ch. 6, § 2.4, for locally compact abelian groups, extends to all symmetric Segal algebras and in part even to pseudosymmetric ones - a result due, essentially, to BURNHAM [1,2], in an even more general formulation.

The foundations having been laid, the results that form the main aim of these lectures can now be established.

In § 10 closed ideals associated with closed subgroups H of G are introduced for symmetric, and in part also for pseudosymmetric, Segal algebras $S^1(G)$, in analogy with the definition given for $L^1(G)$ in § 1. Two theorems are proved which establish a close connection between the existence of approximate units in these ideals and the property P_1 of the subgroup H.

In § 11 the associated ideals are further studied, by two methods; one of these is the method of linear functionals which plays a considerable role later (§ 16).

In § 12 normal subgroups and their associated ideals in symmetric Segal algebras are considered. Here the theorems of § 10, combined with the results of § 11, show their full power, yielding the following corollary, among others: the associated ideal of a closed, normal subgroup H has approximate left (or right) units, bounded in the L^1-norm, if and only if H has the property P_1; moreover, if this is the case, then for a large class of Segal algebras the associated ideal has approximate units with special properties which are of interest even in the abelian case.

In § 13 it is shown that a Segal algebra $S^1(G)$ gives rise, in a natural way, to Segal algebras $S^1(G/H)$ for quotient groups of G, and that various properties of $S^1(G)$ carry over to $S^1(G/H)$. It was proved in Ch. 8, § 4.6 (ii), that the image of a closed ideal of $L^1(G)$ under the natural morphism $L^1(G) \to L^1(G/H)$ is a <u>closed</u> ideal of $L^1(G/H)$, if H has the property P_1; this result is extended here to symmetric and pseudosymmetric Segal algebras.

§ 14 is devoted to the actual construction of closed ideals possessing various kinds of approximate units, in symmetric and pseudosymmetric Segal algebras. Here abelian groups and soluble groups are considered; the general lemmas on approximate units in § 7 play a basic role in this context, jointly with the results in §§ 12 and 13.

In § 15 it is proved that for symmetric Segal algebras on compact groups all (two-sided) ideals have approximate units. Here, of course, some special features appear, and also some analogies with the abelian case. As an application, Segal algebras on semidirect products of compact and locally compact abelian groups are considered and the construction of ideals with various kinds of approximate units in these Segal algebras is discussed, in analogy with the method of § 14.

In § 16 Segal algebras $S^1(G)$ on abelian groups G are studied. It is proved that in this case the bijective correspondence between the closed ideals of $S^1(G)$ and those of $L^1(G)$ established in Ch. 6, § 2.4 (and discussed again in § 9 of these lectures) has the following property: to closed ideals with approximate units in $S^1(G)$ correspond closed ideals with approximate units in $L^1(G)$, and vice-versa. In the proof - which is based on the method of linear functionals, already used in § 11 - the abelian groups structure enters decisively at one particular point; it is an open question whether the result holds for groups that are not abelian (nor compact, the compact case having been settled in § 15). In connection with this result, Wiener-Ditkin sets in abelian groups (Ch. 7, § 4) are discussed, in particular the injection theorem, the proof of which is given here in an improved form.

§ 17 is concerned with the investigation of those closed (two-sided) ideals in $L^1(G)$, for compact groups G and for locally compact abelian groups, which have approximate units of a particular kind corresponding essentially to approximate units in the traditional sense. Here the situation is particularly satisfactory: the structure of such ideals can be determined and the analogy between compact and locally compact abelian groups appears in a lucid way. Finally, some open questions are mentioned which present themselves quite naturally in this context.

These and other questions raised in the course of these lectures, and the further study of Segal algebras, must be left to the reader's own research.

§ 18 contains additions and corrections to the author's monograph mentioned at the beginning.

TABLE OF CONTENTS

§ 1. Certain ideals in L^1-algebras

In the complex Banach algebra $L^1(G)$ - G being a locally compact (l.c.) group - we have the left translation operator L_y and the right translation operator R_y, defined by

(1) $$L_y f(x) = f(y^{-1}x), \quad R_y f(x) = f(xy^{-1})\Delta_G(y^{-1}), \quad x,y \in G,$$

where Δ_G is the Haar modulus of G:

$$\int R_y f(x)dx = \int f(x)dx \qquad f \in L^1(G),$$

dx denoting left Haar measure; we often write $\int f$ for $\int f(x)dx$.

We have

$$L_{ab} = L_a L_b, \quad R_{ab} = R_b R_a \qquad a,b \in G.$$

The mappings $y \to L_y f$, $y \to R_y f$ of G into $L^1(G)$ are continuous; also, the norm in $L^1(G)$ is invariant under L_y, R_y:

(2) $$\|L_y f\|_1 = \|f\|_1, \quad \|R_y f\|_1 = \|f\|_1 \qquad f \in L^1(G), \ y \in G.$$

Let H be a closed subgroup of G. We define

(3) $$D_H L^1(G), \ D'_H L^1(G)$$

as the closed linear subspaces of $L^1(G)$ generated, respectively, by the sets

(4) $$\{f - L_y f \mid f \in L^1(G), y \in H\}, \quad \{f - R_y f \mid f \in L^1(G), y \in H\}.$$

Thus $D_H L^1(G)$, $D'_H L^1(G)$ are the closure in $L^1(G)$ of the set of all linear combinations of functions of the form $f - L_y f$, $f - R_y f$, respectively, f ranging over $L^1(G)$ and y over the subgroup H. We observe:

(5) $$\begin{cases} D_H L^1(G) \text{ is a (closed) right ideal of } L^1(G); \\ D'_H L^1(G) \text{ is a (closed) left ideal of } L^1(G). \end{cases}$$

This follows at once from the relations

(6) $$(L_y f) * g = L_y(f * g), \quad g * (R_y f) = R_y(g * f),$$

where $f * g$ is the convolution of $f, g \in L^1(G)$:

$$f * g(x) = \int f(y) g(y^{-1}x)\, dy.$$

We have

(7) $$D'_H L^1(G) = [D_H L^1(G)]^*,$$

where the star denotes the usual involution in $L^1(G)$:

(8) $$g^*(x) = \overline{g(x^{-1})} . \Delta_G(x^{-1}).$$

To verify (7), we note that

(9) $$R_y f = (L_{y^{-1}}(f^*))^*.$$

Thus $f - R_y f = (f^* - L_{y^{-1}}(f^*))^*$, whence (7) follows via the familiar properties

(10a) $(c_1 g_1 + c_2 g_2)^* = \bar{c}_1 g_1^* + \bar{c}_2 g_2^*$, (10b) $\|g^*\|_1 = \|g\|_1$.

We also recall that

$$(f * g)^* = g^* * f^*.$$

We now prove the following result.

PROPOSITION 1. The equality

(11) $$D'_H L^1(G) = D_H L^1(G)$$

holds if and only if the closed subgroup H is normal.

The <u>proof</u> is based on a lemma. We first recall the notations

(12) $\langle f,\phi\rangle = \int f(x)\overline{\phi(x)}dx \quad f \in L^1(G),\ \phi \in L^\infty(G)$

and

$$\phi \perp f$$

(' ϕ is orthogonal to f '), which means: $\langle f,\phi\rangle = 0$; likewise, for a subset I of $L^1(G)$, $\phi \perp I$ means: $\phi \perp f$ for all $f \in I$. We can now state:

LEMMA 1. Let I be a closed left ideal of $L^1(G)$. If for $f_o \in L^1(G)$ we have $\phi \perp f_o$ for every <u>continuous</u> $\phi \in L^\infty(G)$ such that $\phi \perp I$, then $f_o \in I$.

<u>Proof of Proposition</u> 1. First we note that the continuous bounded functions ϕ orthogonal to $D_H L^1(G)$, $D'_H L^1(G)$, respectively, are those and <u>only</u> those which are constant on each coset H.x, x.H, respectively: this follows readily from the definition of $D_H L^1(G)$, $D'_H L^1(G)$ <u>and</u> the continuity of ϕ.

Now suppose H is normal. Then we apply Lemma 1 to the closed left ideal $D'_H L^1(G)$ and any f_o of the form $f_o = f - L_y f$ ($f \in L^1(G)$, $y \in H$): as the cosets H.x, x.H coincide, we obtain $f_o \in D'_H L^1(G)$, whence

$$D_H L^1(G) \subset D'_H L^1(G),$$

since the linear subspace $D'_H L^1(G)$ is closed. Thus by (7)

$$D_H L^1(G) \subset [D_H L^1(G)]^*;$$

by involution we obtain the opposite inclusion and hence (11).

Conversely, suppose (11) holds. Then every continuous bounded function on G constant on each coset x.H is also constant on each coset H.x, and this cannot happen unless H is normal: indeed, if there is an $x \in G$ and an $\eta \in H$ such that $\eta x \notin x.H$, then there is a function $\dot{\phi} \in \mathcal{K}(G/H)$ such that $\dot{\phi}(\pi_H(\eta x)) = 1$, where π_H is the canonical map of

G onto the quotient space G/H of left cosets y.H ($y \in G$), and $\dot{\phi}(\pi_H(x.H)) = 0$; then $\phi = \dot{\phi} \circ \pi_H$ is a bounded, continuous function on G constant on each coset y.H and such that, for the elements x, η considered, we have $\phi(x) = 0$, $\phi(\eta x) = 1$, i.e. ϕ is not constant on H.x.

<u>COROLLARY of Proposition</u> 1. If H is normal, then $D_H L^1(G)$ is a (closed) two-sided ideal of $L^1(G)$. This now follows on account of (5).

<u>Proof of Lemma</u> 1. The proof has already been given in Ch. 3, § 6.3, but we shall discuss it here again, in a slightly different way, for the sake of later developments.

We will show, using the notation of Lemma 1: if the condition of Lemma 1 holds, then for <u>any</u> ϕ in $L^\infty(G)$ such that $\phi \perp I$ we have $\phi \perp f_o$; this will imply $f_o \in I$. Now observe that for any $g \in L^1(G)$ we also have $g^* * \phi \perp I$, if $\phi \perp I$: this follows from the relation

(13) $$\langle g * f\ ,\ \phi \rangle = \langle f\ ,\ g^* * \phi \rangle \qquad f, g \in L^1(G),\ \phi \in L^\infty(G),$$

since by assumption I is a left ideal. The function $g^* * \phi \in L^\infty(G)$ is continuous, thus by hypothesis we have $g^* * \phi \perp f_o$; using (13) again (with f_o in the place of f), we get

(14) $$\langle g * f_o\ ,\ \phi \rangle = 0 \qquad \text{for all } g \in L^1(G).$$

Now we recall that $L^1(G)$ has 'approximate left units': given $f \in L^1(G)$ and any $\varepsilon > 0$, there is a $u \in L^1(G)$ such that

(15) $$\| u * f - f \|_1 < \varepsilon .$$

Applying this to $f = f_o$ and putting $g = u$ in (14), we obtain $\langle f_o, \phi \rangle =$ $= 0$, that is, $\phi \perp f_o$. Thus $f_o \in I$, as was to be proved. The proof given in Ch. 3, § 6.3 does not make use of approximate left units.

There is another characterization of $D_H L^1(G)$, $D'_H L^1(G)$ which we shall now discuss.

Let us first consider the case of a normal subgroup H. Then the quotient group G/H carries a left Haar measure and there is a mapping T_H of $L^1(G)$ onto $L^1(G/H)$ defined by

(16) $$T_H f(\dot{x}) = \int_H f(x\xi)d\xi \qquad f \in L^1(G),\ \dot{x} = \pi_H(x) \in G/H,$$

where $d\xi$ is a (fixed) left Haar measure on H. Moreover, $T_H f$ is in $L^1(G/H)$ and the left Haar measure $d\dot{x}$ on G/H can be so chosen that

(17) $$\int_{G/H} T_H f(\dot{x})d\dot{x} = \int_G f(x)dx \qquad \text{for all } f \in L^1(G)$$

(cf. Ch. 3, § 3.3 (i) and § 4). From (17) it follows that

(18) $$\|T_H f\|_1 \leqslant \|f\|_1,$$

where the norm on the left refers to $L^1(G/H)$. The mapping T_H of $L^1(G)$ onto $L^1(G/H)$ is not only linear, but also a star-homomorphism (cf. Ch. 3, § 5.3), i.e.

(19) $$T_H(f * g) = T_H f * T_H g,$$

where the convolution on the right is that in $L^1(G/H)$, and

(20) $$T_H(f^*) = (T_H f)^*,$$

where the star on the right means involution in $L^1(G/H)$.

Let H be normal and denote by $J^1(G,H)$ the kernel of T_H:

(21) $$J^1(G,H) = \{f \mid f \in L^1(G),\ T_H f = 0\}.$$

Then $J^1(G,H)$ is closed, by (18), and is a two-sided ideal in $L^1(G)$ (cf. (19)); also, (20) implies that

(22) $$[J^1(G,H)]^* = J^1(G,H).$$

It is easy to see that, if H is normal,

(23) $$D_H L^1(G) \subset J^1(G,H),$$

and Lemma 1 can be used to show that for <u>normal</u> H

(24) $$D_H L^1(G) = J^1(G,H)$$

(cf. Ch. 3, § 6.4). Incidentally, (22) now yields another proof of the fact that, if H is normal, then $[D_H L^1(G)]^* = D_H L^1(G)$ (cf. (7), (11)).

Now we consider the case of an arbitrary (closed) subgroup $H \subset G$. In this case the quotient <u>space</u> G/H still carries a 'quasi-invariant' measure and the following holds (cf. Ch. 8, § 1). There exists a strictly positive continuous function q on G satisfying

(25) $$q(x\xi) = q(x)\Delta_H(\xi)/\Delta_G(\xi) \text{ for all } x \in G, \ \xi \in H.$$

Let

(26) $$T_{H,q}f(\dot{x}) = \int_H \frac{f(x\xi)}{q(x\xi)}d\xi \qquad f \in L^1(G), \ \dot{x} = \pi_H(x) \in G/H;$$

then we have (cf. Ch. 8, §§ 2.1 - 2.3): there is a quasi-invariant measure $d_q x$ on G/H such that

$$T_{H,q}f \text{ is in } L^1(G/H) = L^1(G/H, d_q\dot{x})$$

and

(27) $$\int_{G/H} T_{H,q}f(\dot{x})d_q\dot{x} = \int_G f(x)dx \text{ for all } f \in L^1(G).$$

$T_{H,q}$ is a linear mapping of $L^1(G)$ <u>onto</u> $L^1(G/H)$; moreover, the analogue of (18) holds:

(28) $$\|T_{H,q}f\|_1 \leqslant \|f\|_1 \qquad f \in L^1(G).$$

Denote the kernel of $T_{H,q}$ again by $J^1(G,H)$:

(29) $$J^1(G,H) = \{f \mid f \in L^1(G), \ T_{H,q}f = 0\}.$$

This is a closed linear subspace of $L^1(G)$ (cf. (28)) and is, in fact, independent of a particular choice of the function q. The relation

(30) $$T_{H,q}R_y f = T_{H,q} f \qquad \text{if } y \in H,$$

(proved via (26)) shows at once that

(31) $$D'_H L^1(G) \subset J^1(G,H)$$

and hence, by (7), we also have

(32) $$D_H L^1(G) \subset [J^1(G,H)]^*.$$

Moreover, by means of Lemma 1 it can be shown that, in fact,

(33) $$D'_H L^1(G) = J^1(G,H)$$

(cf. Ch. 8, § 2.5); thus (7) becomes

(34) $$D_H L^1(G) = [J^1(G,H)]^*.$$

If H is normal, then for the function q in (25) we may choose the constant 1, so that $T_{H,q}$, defined by (26), reduces to T_H, as defined by (16) (the quasi-invariant measure $d_q\dot{x}$ is then simply the left Haar measure $d\dot{x}$ on G/H); thus the definition (29) of $J^1(G,H)$ is an extension of (21). Also the relations (33) and (34) coalesce if H is normal.

As the relations (34) and (33) show, the subspaces $D_H L^1(G)$ and $D'_H L^1(G)$ introduced in this section could be <u>defined</u> in terms of $J^1(G,H)$, and Proposition 1 could then be stated in terms of $J^1(G,H)$ alone. But, as the reader may convince himself, the procedure followed here leads to a simpler proof of Proposition 1; besides, $D_H L^1(G)$ and $D'_H L^1(G)$ are defined in a more elementary way than $J^1(G,H)$ if the subgroup H is not normal.

§ 2. The property P_1; applications

We say that a l.c. group G has the property P_1 (Ch. 8, § 3.1), or is a P_1-group, if, given any compact set $E \subset G$ and $\varepsilon > 0$, there is a function $s \in L^1(G)$ such that $s \geqslant 0$, $\int s = 1$, and

$$(1) \qquad \| L_y s - s \|_1 < \varepsilon \qquad \text{for all } y \in E.$$

The class of P_1-groups is discussed in Ch. 8. §§ 3.3 and 7.

P_1 implies an apparently stronger property: if we put

$$(2) \qquad s_2 = s * s^*,$$

then s_2 satisfies $s_2 \geqslant 0$, $\int s_2 = 1$ and

$$(3) \qquad \| L_y s_2 - s_2 \|_1 < \varepsilon, \quad \| R_y s_2 - s_2 \|_1 < \varepsilon \quad \text{for all } y \in E.$$

For, we have $L_y s_2 - s_2 = (L_y s - s) * s^*$ and

$$R_y s_2 - s_2 = s * (L_{y^{-1}} s - s)^*$$

(cf. § 1, (6), (9)); thus e.g.

$$\| R_y s_2 - s_2 \|_1 \leqslant 1.\| L_{y^{-1}} s - s \|_1 = \| s - L_y s \|_1$$

(cf. § 1, (10b)).

For the subject-matter of these lectures the following applications of the property P_1 will be basic.

Let H be a closed subgroup of G. If H has the property P_1, then for every $f \in L^1(G)$ the relation

$$(4) \qquad \inf \| \Sigma_n c_n R_{\xi_n} f \|_1 = \| T_{H,q} f \|_1$$

holds, where the infimum is taken over all (finite) sums with elements $\xi_n \in H$ and coefficients $c_n > 0$ satisfying

(5) $$\Sigma_n c_n = 1.$$

For a fuller discussion and the proof - which is entirely by the methods of classical analysis, extended to groups - see Ch. 8, §§ 4.1 - 4.2 (let us mention that the operator $A_\xi = R_{\xi^{-1}}$ is used there, which of course amounts to the same, since $\xi_n \in H$ is arbitrary).

From (4) we can obtain a corresponding relation for left translations:

(6) $$\inf \|\Sigma_n c_n L_{\xi_n} f\|_1 = \|T_{H,q}(f^*)\|_1 \qquad f \in L^1(G),$$

where the infimum is taken in the same way as in (4); the star on the right means involution in $L^1(G)$ (§ 1, (8)). To obtain (6), we apply (4) to the function f^* and then use the formulae (9) and (10a,b) of § 1.

We now introduce, once for all, the notations $\mathcal{L}_H$, $\mathcal{R}_H$ for the convex hull of all operators L_ξ, R_ξ ($\xi \in H$), respectively: $\mathcal{L}_H$ and $\mathcal{R}_H$ consist of all operators L, R, respectively, of the form

$$L = \Sigma_n c_n L_{\xi_n}, \quad R = \Sigma_n c_n R_{\xi_n},$$

with

$$c_n > 0, \quad \Sigma_n c_n = 1, \quad \xi_n \in H.$$

In the case of a <u>normal</u> P_1-subgroup H, the formulae (4) and (6) reduce to

(7) $$\inf_{L \in \mathcal{L}_H} \|Lf\|_1 = \|T_H f\|_1 \qquad f \in L^1(G),$$

(8) $$\inf_{R \in \mathcal{R}_H} \|Rf\|_1 = \|T_H f\|_1 \qquad f \in L^1(G),$$

i.e. on the right only $T_H f$ appears, even in the case of $\mathcal{L}_H$: this is so because of § 1, (20).

From (6) and (4) we have in particular, for <u>any</u> P_1-subgroup H:

(9) If $f \in [J^1(G,H)]^*$, then $\inf_{L \in \mathcal{L}_H} \|Lf\|_1 = 0$.

(10) If $f \in J^1(G,H)$, then $\inf_{R \in \mathcal{R}_H} \|Rf\|_1 = 0$.

REMARK 1. The inclusion § 1, (32) shows that in (9) we may replace $[J^1(G,H)]^*$ by $D_H L^1(G)$; likewise § 1, (31) shows that in (10) we may replace $J^1(G,H)$ by $D_H' L^1(G)$. The equalities § 1, (34) and (33) show that this is, in fact, no loss of generality.

We can extend (9) and (10) as follows.

PROPOSITION 1. Let H be a closed P_1-subgroup of G. Then, given finitely many functions in $D_H L^1(G)$, say $(f_j)_{1 \leqslant j \leqslant N}$, and $\varepsilon > 0$, there is an $L \in \mathcal{L}_H$ such that

(11) $$\|Lf_j\|_1 < \varepsilon \qquad \text{for } j = 1,\ldots,N.$$

Likewise, if functions $(f_j)_{1 \leqslant j \leqslant N}$ in $D_H' L^1(G)$ and $\varepsilon > 0$ are given, there is an $R \in \mathcal{R}_H$ such that

(12) $$\|Rf_j\|_1 < \varepsilon \qquad \text{for } j = 1,\ldots,N.$$

Proof. Consider e.g. (11): for N = 1, this holds by (9) (cf. also Remark 1). Suppose that (11) holds for N-1 in place of N, if N > 1: thus there is an operator $A_1 \in \mathcal{L}_H$ such that

$$\|A_1 f_j\|_1 < \varepsilon \quad \text{for } 1 \leqslant j \leqslant N-1.$$

Now consider $A_1 f_N$ and observe that

$$A_1 f_N \in D_H L^1(G)$$

[for, if $\xi \in H$ and $f \in D_H L^1(G)$, then $L_\xi f = f - (f - L_\xi f)$ lies in $D_H L^1(G)$, thus Af lies in $D_H L^1(G)$ for every $A \in \mathcal{L}_H$]. By (9), applied to $A_1 f_N$, there is an operator $A_2 \in \mathcal{L}_H$ such that

$$\|A_2(A_1 f_N)\|_1 < \varepsilon.$$

Now put $L = A_2A_1$: then $L \in \mathcal{L}_H$ and $\|Lf_j\|_1 < \varepsilon$ for <u>all</u> j, $1 \leqslant j \leqslant n$ [for $1 \leqslant j \leqslant N-1$ note that $\|A_2(A_1f_j)\|_1 \leqslant \|A_1f_j\|_1$]. Thus (11) is proved. (12) follows from (11) by involution (cf. § 1, (7), (9)).

We shall see applications of Proposition 1 later (§ 10). Let us observe here that the method by which Proposition 1 was proved will be used again, in a somewhat different context, in the proof of Lemma 1 in § 7; the method goes back to Ch. 8, § 4.5.

§ 3. Some properties equivalent to P_1. Relativization

The property P_1 belongs entirely to classical analysis. It is, however, closely related to another property of a quite different nature. We say that a l.c. group G has the *property* (M) if there exists a left invariant mean on $L^\infty(G)$. *The properties* (M) *and* P_1 *are equivalent* (Ch. 8, §§ 5, 6).

The proof that P_1 implies (M) is straightforward, being based on a compactness argument in the dual of $L^\infty(G)$: see Ch. 8, § 5.4; let us remark that the functionals considered there may be written simply in the form

$$\phi \to \int \phi(t) s_2(t)dt,$$

with $s_2 = s * s^*$ (cf. § 2, (2), (3)), and that the use of s_2 rather than s yields a somewhat stronger property than (M) (cf. loc. cit.).

The proof that, conversely, (M) implies P_1 is more recondite, being based on - among other things - a theorem of Glicksberg's (Ch. 8, § 6).

For the sake of later applications, we now introduce the following definition.

DEFINITION. Let H be a closed subgroup of the l.c. group G. We say that the pair (G,H) has the *property* $P_*(G,H)$ $[P_1(G,H)]$ if, given any finite [compact] set $E \subset H$ and $\varepsilon > 0$, there is a function $s \geqslant L^1(G)$ such that $s \geqslant 0$, $\int s = 1$ and

$$\|L_y s - s\|_1 < \varepsilon \quad \text{for all } y \in E.$$

For H = G, $P_1(G) = P_1(G,G)$ is simply the familiar property P_1 for

G; $P_1(G,H)$ is a <u>relativization</u> of $P_1(G)$ in the sense explained in Ch. 4, § 5. This relativization, together with the introduction of $P_*(G,H)$ - which is simply a weaker form of $P_1(G,H)$ - will be useful later (§ 10). Here we shall prove the following result.

PROPOSITION 1. Let H be a closed subgroup of the l.c. group G. If the property $P_*(G,H)$ holds, then H has the property P_1. Conversely, if H has the property P_1, then $P_1(G,H)$ holds.

<u>Proof</u>. We shall show that $P_*(G,H)$ implies the property (M) for H, which is equivalent to $P_1(H)$. Let β be a Bruhat function for H on G (cf. Ch. 8, § 1.9), $d\xi$ a left Haar measure on H, and define for $\phi \in L^\infty(H)$ the function $\tau_H\phi \in L^\infty(G)$ by

(1) $$\tau_H\phi(x) = \int_H \phi(\xi)\beta(x^{-1}\xi)\, d\xi \qquad x \in G.$$

$\tau_H\phi$ is, in fact, continuous on G (cf. Ch. 8, § 1.9, Remark, and Ch. 3, § 3.2, Lemma) and

(2) $$\|\tau_H\phi\|_\infty \leqslant \|\phi\|_{L^\infty(H)} .$$

Also

(3) $$\tau_H L_y\phi = L_y\tau_H\phi \qquad \text{if } y \in H.$$

Now we shall apply a 'relativization' of the argument in Ch. 8, § 5.4. Denote by E any finite subset of H, let $\varepsilon > 0$, and let $A(E,\varepsilon)$ be the set of all functionals m on $L^\infty(H)$ of the form

$$m(\phi) = \int_G \tau_H\phi(x).s(x)dx \qquad \phi \in L^\infty(H),$$

where s ranges over the subset of $L^1(G)$ given by

(4) $$\{s \mid s \in L^1(G),\ s \geqslant 0,\ \int s = 1,\ \|L_y s - s\|_1 < \varepsilon \text{ for all } y \in E\},$$

which is not empty, by the hypothesis $P_*(G,H)$. Thus $A(E,\varepsilon)$ is a non-empty subset of the unit ball of the dual $L^{\infty}(G)'$, since by (2)

$$|m(\phi)| \leqslant \|\tau_H\phi\|_{\infty} \leqslant \|\phi\|_{L^{\infty}(H)} ;$$

also, by using (2), (3), (4) and the obvious relation

$$\|L_{a^{-1}}s - s\|_1 = \|L_a s - s\|_1$$

we see that $m \in A(E,\varepsilon)$ satisfies

$$(5) \qquad |m(L_a\phi - \phi)| \leqslant \|\phi\|_{L^{\infty}(H)}\cdot\varepsilon \qquad \text{for all } a \in E.$$

Each functional m is a mean on $L^{\infty}(H)$ and (5) says that m gets 'more and more nearly invariant' as $E \subset H$ gets 'bigger' and $\varepsilon > 0$ smaller. Now let $B(E,\varepsilon)$ be the closure of $A(E,\varepsilon)$ in $L^{\infty}(G)'$ in the topology $\sigma((L^{\infty})',L^{\infty})$; thus by continuity each $m \in B(E,\varepsilon)$ also has these properties: m is a mean on $L^{\infty}(H)$ and satisfies (5). The sets $B(E,\varepsilon)$ are compact (for they are closed subsets of the unit ball of $L^{\infty}(H)'$ which is compact in the topology considered) and the intersection of any finite number of them is non-empty (since this is obviously true for the sets $A(E,\varepsilon)$). Thus the intersection

$$\bigcap_{E,\varepsilon} B(E,\varepsilon) \quad (E \subset H \text{ finite, } \varepsilon > 0)$$

is non-empty. Any functional in this intersection is a left invariant mean on $L^{\infty}(H)$, i.e. H has the property (M) and hence the property P_1.

The converse can be shown more directly. Let any <u>compact</u> set $K \subset H$ and $\varepsilon > 0$ be given. By hypothesis there is an $s_H \in L^1(H)$ such that

$$\|L_y s_H - s_H\|_{L^1(H)} < \varepsilon \qquad \text{for all } y \in K.$$

Now choose any $u \in \mathcal{K}_+(G)$ such that $\int_G u = 1$ and put

$$s(x) = \int_H s_H(\xi)u(\xi^{-1}x)d\xi \qquad x \in G.$$

Then s is clearly continuous, $s \geqslant 0$, and

$$\int_G s = \int_H s_H . \int_G u = 1.$$

Also

$$\|L_\eta s - s\|_1 = \int_G \left| \int_H s_H(\xi)\{u(\xi^{-1}\eta^{-1}x) - u(\xi^{-1}x)\}d\xi \right| dx$$

$$= \int_G \left| \int_H \{s_H(\eta^{-1}\xi) - s_H(\xi)\}u(\xi^{-1}x)d\xi \right| dx$$

$$\leqslant \|L_\eta s_H - s_H\|_{L^1(H)} . \int_G u(x)dx$$

$$< \varepsilon \quad \text{for all } \eta \in K.$$

Thus $P_1(G,H)$ holds and the proof is complete.

It should be observed that even in the case H = G the proof that $P_*(G)$ implies $P_1(G)$ - and hence is equivalent to $P_1(G)$ - is not trivial: it depends on the property (M), just as in the 'relativized' case; no more direct proof seems available; of course, when H = G, the Bruhat function is not needed. The case H = G was discussed in REITER [2].

§ 4. Segal algebras

Let G be an arbitrary l.c. group. A linear subspace of $L^1(G)$ is said to be a Segal algebra, and denoted by $S^1(G)$, if it satisfies the following conditions.

(S_o) $S^1(G)$ is dense in $L^1(G)$.

(S_1) $S^1(G)$ is a Banach space under some norm $\|.\|_S$, and

$$\|f\|_S \geq \|f\|_1 \qquad \text{for all } f \in S^1(G). \tag{1}$$

(S_2) $S^1(G)$ is left invariant ($f \in S^1(G) \Rightarrow L_y f \in S^1(G)$ for all $y \in G$) and for each $f \in S^1(G)$ the mapping $y \to L_y f$ of G into $S^1(G)$ is continuous.

The continuity already holds if it holds for $y = e$; this means: given $f \in S^1(G)$ and $\varepsilon > 0$, there is a neighbourhood (nd.) U of e such that

$$\|L_y f - f\|_S < \varepsilon \qquad \text{for all } y \in U. \tag{2}$$

(S_3) The Segal norm is left invariant: $\|L_y f\|_S = \|f\|_S$ for all $f \in S^1(G)$ and all $y \in G$.

(S_1) says only that $S^1(G)$ is a Banach space, but it will shortly be shown that $S^1(G)$ is, in fact, a Banach algebra under convolution (cf. Proposition 1 below). Also, it would be enough to require in the place of (1) the existence of a constant C such that

$$\|f\|_1 \leq C.\|f\|_S \qquad \text{for all } f \in S^1(G);$$

but the assumption $C = 1$ is no loss of generality: if $C > 1$, we can replace the given norm $\|f\|_S$ by the equivalent norm $C.\|f\|_S$ (this even works for $S^1(G)$ as a Banach algebra). On the other hand, if G is abelian, then (S_1) can be replaced by the condition that $S^1(G)$ be a Banach algebra under the norm $\|.\|_S$; this case was treated in Ch. 6, § 2.

We now list some other conditions that will be used in the sequel and which a Segal algebra may, or may not, satisfy. First we have the 'right-hand' versions of (S_2) and (S_3):

(S_2') $S^1(G)$ is right invariant ($f \in S^1(G) \Rightarrow R_y f \in S^1(G)$ for all $y \in G$) and for each $f \in S^1(G)$ the mapping $y \to R_y f$ of G into $S^1(G)$ is continuous, which is equivalent to (2), with L_y replaced by R_y.

(S_3') The Segal norm is right invariant: $\|R_y f\|_S = \|f\|_S$ for all $f \in S^1(G)$ and all $y \in G$.

We call a Segal algebra symmetric if it satisfies (S_2') and (S_3'). Note that the symmetric Segal algebras include all Segal algebras on l.c. abelian groups.

The condition (S_3') is rather restrictive; we therefore introduce another condition as follows.

(S^O) $S^1(G)$ contains functions $u \geqslant 0$ with $\int u = 1$ and arbitrarily small support: given any nd. U of e in G, there is a $u \in S^1(G)$ such that $u \geqslant 0$, $\int u = 1$ and Supp $u \subset U$.

The question whether every Segal algebra satisfies (S^O) remains to be settled; on the other hand, it will be shown below that (S^O) implies (S_O).

We shall call a Segal algebra pseudosymmetric if it satisfies (S_2') and (S^O). Examples of Segal algebras that are pseudosymmetric, but not symmetric, are readily given (cf. § 5).

At times we shall also use another condition, stronger than symmetry: we call a Segal algebra $S^1(G)$ star-symmetric if $S^1(G)$ is stable under the involution $f \to f^*$ of $L^1(G)$ and if this involution is an isometry of $S^1(G)$: $\|f^*\|_S = \|f\|_S$ for all $f \in S^1(G)$.

Clearly star-symmetry implies symmetry (cf. § 1, (9)), but the converse is an open question (even for abelian groups!).

In the study of Segal algebras, integration of functions with values in a Banach space is frequently used; for a summary of this theory, notation and references, see Ch. 3, § 2. We state here the following simple lemma for later use.

LEMMA 0. Let X be a l.c. space, μ a positive measure on X, and B a Banach space with norm $\|.\|$.

(i) If $\underline{f}$ is a bounded, continuous function on X with values in B, then for every $h \in L^1(X,\mu)$ the function $h.\underline{f}$ belongs to $L^1_B(X,\mu)$.

(ii) If $\underline{f}$ is any continuous function on X with values in B, then for every $h \in L^1(X,\mu)$ with compact support the function $h.\underline{f}$ belongs to $L^1_B(X,\mu)$.

Proof. (i) Let $\|f(x)\| \leq C$ for all $x \in X$. Then, if $k \in \mathcal{K}(X)$ is such that $N_1(h-k) < \varepsilon$, we have $N_1(h.\underline{f} - k.\underline{f}) < C.\varepsilon$.

(ii) can be reduced to (i): for, let $k_1 \in \mathcal{K}(X)$ be such that $k_1(x) = 1$ for all $x \in \operatorname{Supp} h$; then $h.\underline{f} = h.(k_1.\underline{f})$.

By means of this lemma, we can now obtain some properties of Segal algebras.

LEMMA 1.

(i) For every $f \in S^1(G)$ and arbitrary $h \in L^1(G)$ the vector-valued integral $\int h(y)L_y f dy$ exists, as an element of $S^1(G)$, and

$$\int h(y)L_y f dy = h * f. \tag{3}$$

(ii) If $S^1(G)$ satisfies (S_2'), then for every $f \in S^1(G)$ and for every $h \in L^1(G)$ with compact support the vector-valued integral $\int h(y)R_y f dy$ exists, as an element of $S^1(G)$, and

$$\int h(y)R_y f dy = f * h. \tag{4}$$

(iii) If $S^1(G)$ is symmetric, then for every $f \in S^1(G)$ and arbitrary $h \in L^1(G)$ the vector-valued integral $\int h(y)R_y f dy$ exists, as an element of $S^1(G)$, and satisfies (4).

Proof.

(i) See Lemma 0, part (i), and the first half of the proof in Ch. 6, § 2.2 (ii).

(ii) This follows from Lemma 0, part (ii), and an argument entirely analogous to that in Ch. 6, § 2.2 (ii), with L_y replaced by R_y.

(iii) If $S^1(G)$ is symmetric, we can apply Lemma 0, part (i), to the function $y \to R_y f$.

REMARK 1. The analogy between the relations (3) and (4) should be noted (cf. also (5) and (6) below). They embody the principle of symmetry for Segal algebras: if $S^1(G)$ is symmetric, we can simply interchange, in a proof, L_y and R_y and the order of convolution, to obtain a 'symmetric' result. This already appears in § 1, (6).

We now have the following result.

PROPOSITION 1.

(i) Any Segal algebra $S^1(G)$ is a left ideal of $L^1(G)$ and

(5) $$\|h * f\|_S \leqslant \|h\|_1 . \|f\|_S \qquad f \in S^1(G),\ h \in L^1(G).$$

In particular, $S^1(G)$ is a Banach algebra under $\|.\|_S$.

(ii) If $S^1(G)$ is symmetric, then $S^1(G)$ is also a right ideal of $L^1(G)$ and

(6) $$\|f * h\|_S \leqslant \|h\|_1 . \|f\|_S \qquad f \in S^1(G),\ h \in L^1(G).$$

Proof.

(i) This follows from Lemma 1, part (i), by an application of (S_3) (cf. the second half of the proof in Ch. 6, § 2.2 (ii)); if $h \in S^1(G)$, then (5) implies $\|h * f\|_S \leqslant \|h\|_S . \|f\|_S$ (cf. (1)).

(ii) This follows by symmetry.

We can now show that (S^{o}) implies (S_{o}): by relation (5) (the proof of which depends only on (S_1), (S_2), (S_3)) we have, if $u \in S^1(G)$, $f * u \in S^1(G)$ for any $f \in L^1(G)$ and $f * u \to f$ in $L^1(G)$ if $u \geqslant 0$, $\int u = 1$ and Supp $u \to \{e\}$.

Proposition 1 may be generalized.

PROPOSITION 2. Let μ be a bounded, complex measure on G. Then $\mu * f$ is in $S^1(G)$ for any $f \in S^1(G)$ and

$$(7) \qquad \|\mu * f\|_S \leqslant \|\mu\|.\|f\|_S .$$

If $S^1(G)$ is symmetric, then also $f * \mu \in S^1(G)$ and

$$(8) \qquad \|f * \mu\|_S \leqslant \|\mu\|.\|f\|_S .$$

The proof is entirely analogous to that of Proposition 1; for the sake of completeness, we discuss it here.

(i) Let us first make the brief indications in Ch. 3, § 2.3 a little more explicit and more general. Let X be a l.c. space and μ any bounded complex measure on X (loc. cit. only positive measures are considered); let $|\mu|$ be the positive measure associated with μ (Ch. 3, § 2.1 (viii)). Let B be a Banach space (with norm $\|.\|$). We then define $\mathcal{L}^1_B(X,\mu)$ and $L^1_B(X,\mu)$ to be the same as $\mathcal{L}^1_B(X,|\mu|)$ and $L^1_B(X,|\mu|)$, respectively; thus

$$(9) \quad N_1(\underline{f}) = \int^{\times} \|\underline{f}(x)\| \, d|\mu|(x) = \int \|\underline{f}(x)\| \, d|\mu|(x) \qquad \underline{f} \in \mathcal{L}^1_B(X,\mu).$$

Let us now discuss the definition of the vector integral $\int \underline{f} d\mu$. Consider first functions $\underline{f}$ on X of the form

$$(10) \qquad \underline{f}(x) = \Sigma_n f_n(x).\underline{a}_n \qquad x \in X,$$

where $\underline{a}_n \in B$ and $f_n \in \mathcal{K}(X)$; these functions form a linear subspace $\mathcal{K}^o_B$ of $\mathcal{K}_B(X)$. We define

(11) $$\int \underline{f}(x)\, d\mu(x) = \Sigma_n \int f_n(x)\, d\mu(x).\underline{a}_n \quad \text{for } \underline{f} \in \mathcal{K}_B^{\circ}.$$

To show that the left-hand side of (11) is independent of a particular representation (10) of $\underline{f} \in \mathcal{K}_B^{\circ}$, it is enough to note that (11) implies

(12) $$\langle \int \underline{f}(x)\, d\mu(x), \underline{z}' \rangle = \int \langle \underline{f}(x), \underline{z}' \rangle\, d\mu(x) \text{ for all } \underline{z}' \in B',$$

where B' is the dual of B; here $\langle \underline{z}, \underline{z}' \rangle$ is the value of $\underline{z}'$ at $\underline{z} \in B$.

Next we observe that for every $\underline{z} \in B$ there exists a $\underline{z}' \in B'$ such that $\|\underline{z}'\|_{B'} = 1$ and $\langle \underline{z}, \underline{z}' \rangle = \|\underline{z}\|$. Applying this to $\underline{z} = \int \underline{f}(x)\, d\mu(x)$, we obtain from (12)

$$\|\int \underline{f}(x)\, d\mu(x)\| \leqslant \int \|\underline{f}(x)\|\, d|\mu|(x)$$

or, in shorter notation,

(13) $$\|\int \underline{f}\, d\mu\| \leqslant N_1(\underline{f}).$$

Incidentally, the same method is used for positive measures in the classical case $B = \mathbb{C}$ (Ch. 3, § 2.1 (iv)); in the general case above, the method of linear functionals plays an essential role.

We shall show below that

(14) $\mathcal{K}_B^{\circ}$ is dense in $\mathcal{K}_B(X)$ (with the norm (9)).

Thus, in view of (13), the integral can be uniquely extended from $\mathcal{K}_B^{\circ}$ to $\mathcal{K}_B(X)$ and hence to $\mathcal{L}_B^1(X,\mu)$, and the fundamental inequality (13) holds for all $\underline{f} \in \mathcal{L}_B^1(X,\mu)$; likewise for (12).

To prove (14), let $\underline{f} \in \mathcal{K}_B(X)$ be given. Choose $g \in \mathcal{K}_+(X)$ such that $g(x) = 1$ for $x \in \operatorname{Supp} \underline{f}$; put $K = \operatorname{Supp} g$, a compact set. Now, given $\varepsilon > 0$, there are finitely many points $a_n \in K$, say $1 \leqslant n \leqslant N$, and open nds. U_n of a_n such that $\|\underline{f}(x) - \underline{f}(a_n)\| < \varepsilon$ for $x \in U_n$. Next there are N functions $e_n \in \mathcal{K}_+(X)$ such that $\operatorname{Supp} e_n \subset U_n$ and $\Sigma_n e_n(x) = 1$ for $x \in K$ (this is rather simple to show, cf. e.g. Ch. 2, § 3.1, Remark). Then we have for each n

(15) $\|e_n(x).(\underline{f}(x) - \underline{f}(a_n))\| \leqslant e_n(x).\varepsilon$ for <u>all</u> $x \in X$

and also

(16) $\underline{f} - g.\Sigma_n e_n.\underline{f}(a_n) = g.\Sigma_n e_n.(\underline{f} - \underline{f}(a_n))$.

Put $\underline{k} = g.\Sigma_n e_n.\underline{f}(a_n)$; then $\underline{k} \in \mathcal{K}_B^o$ and (15) and (16) yield

$$\|\underline{f}(x) - \underline{k}(x)\| \leqslant g(x).\Sigma_n e_n(x).\varepsilon = g(x).\varepsilon ,$$

whence $N_1(\underline{f}-\underline{k}) \leqslant N_1(g).\varepsilon$, which proves (14), since g is independent of ε.

BOURBAKI treats vector integration in still greater generality and with a slightly different approach (for references see Ch. 3, § 2.3).

(ii) Now consider $B = S^1(G)$ and $\mu \in M^1(G)$. Since $y \to L_y f$, for $f \in S^1(G)$, is a bounded, continuous function on G with values in B, and μ is a bounded (complex) measure, we have: $y \to L_y f$ is in $\mathcal{L}_B^1(X,\mu)$; thus $\int L_y f\, d\mu(y)$ exists, as an element of $B = S^1(G)$, and (cf. (13))

$$\|\int L_y f\, d\mu(y)\|_S \leqslant \int \|L_y f\|_S\, d|\mu|(y) = \|\mu\|.\|f\|_S .$$

(iii) It remains to show that

(17) $$\int L_y f\, d\mu(y) = \mu * f.$$

Now the integral satisfies (12); thus, denoting the left-hand side of (17) by $f_1 \in S^1(G)$, we have, in particular,

(18) $\int f_1(x)\phi(x)dx = \int\{\int L_y f(x)\phi(x)dx\}\, d\mu(y)$ for each $\phi \in L^\infty(G)$.

To make matters as simple as possible, we now take ϕ in $\mathcal{K}(G)$. Then the equality

(19) $\int\{\int f(y^{-1}x)\phi(x)dx\}\, d\mu(y) = \int\{\int f(y^{-1}x)\, d\mu(y)\}\, \phi(x)dx$

is elementary if $f \in \mathcal{K}(G)$ and follows for all $f \in \mathcal{L}^1(G)$ by continuity. Now (18), (19) show - since $\phi \in \mathcal{K}(G)$ is arbitrary - that f_1 and $\mu * f$ represent the same element of $L^1(G)$, as their difference has $L^1(G)$-norm zero (cf. Ch. 3, § 2.3 (vii)).

(iv) The proof for $\int R_y f\, d\mu(y)$ follows by the principle of symmetry, for symmetric $S^1(G)$ (cf. Remark 1 above).

A generalization of Segal algebras has been given by CIGLER [1], where a proof of Proposition 2, for abelian groups, will also be found (p. 276). Another generalization was given by BURNHAM [1,2].

In these lectures we shall only consider Segal algebras as defined in this section. The most important Segal algebras to be studied in the sequel are the symmetric ones which show the most analogies with the commutative Segal algebras. But several of the results to be proved also hold for pseudosymmetric Segal algebras, and certain basic results established in § 10 hold for Segal algebras in general. It should also be observed that there are examples of Segal algebras of which we do not know whether they are symmetric or pseudosymmetric (cf. § 5, example (iii)).

The historical development of the study of Segal algebras in the commutative case is described in Ch. 6, § 2. For a compact group G, K(G) - with the uniform norm - and $L^2(G)$ are symmetric Segal algebras; these are already implicit in the classical paper of Peter-Weyl which appeared in 1927 in the Mathematische Annalen.

§ 5. Examples

(i) The continuous functions in $L^1(G)$ that vanish at infinity form a Segal algebra, the norm being defined by $\|f\|_S = \|f\|_1 + \|f\|_\infty$.

(ii) For $1 < p < \infty$ the intersection $L^1(G) \cap L^p(G)$ is a Segal algebra, with norm $\|f\|_S = \|f\|_1 + \|f\|_p$.

The examples (i) and (ii) are pseudosymmetric Segal algebras; they are symmetric if and only if G is unimodular.

(iii) Let G be a l.c. group containing a discrete subgroup Γ such that the quotient space G/Γ has finite invariant measure. Consider the (complex-valued) functions f on G such that

$$(1) \qquad N(f) = \sup_{x\in G} \Sigma_{\gamma\in\Gamma} |f(x\gamma)| < \infty$$

Clearly these functions form a Banach space containing $\mathcal{K}(G)$; the closure of $\mathcal{K}(G)$ in this space is a Segal algebra, with $\|f\|_S = N(f)$ (cf. also Ch. 6, § 2.6). This Segal algebra consists, as is readily seen, of all continuous functions f on G with $N(f) < \infty$ that satisfy the following condition:

(2) given $\varepsilon > 0$, there is a compact set $E \subset G$ such that $N(f.\phi_{\complement E}) < \varepsilon$.

Here $\phi_{\complement E}$ is the characteristic function of the complement of E. Whether this Segal algebra satisfies the condition (S_2') of § 4, is an open question.

(iv) Let in (iii) G/Γ be compact. Then we put

$$(3) \qquad N(f) = \sup_{x,y\in G} \Sigma_{\gamma\in\Gamma} |f(x\gamma y)| .$$

The norm (3) yields, in the same way as (1), a Segal algebra consisting of the continuous functions on G with $N(f) < \infty$ and satisfying (2). We remark that, in showing that the norm (3) is finite for all $f \in \mathcal{K}(G)$, the compactness of G/Γ is used in an essential way. The proof that $y \to L_y f$ is continuous for $f \in \mathcal{K}(G)$, with the norm (3), is quite analogous to that for example (iii). This Segal algebra is symmetric (even star-symmetric, since G is necessarily unimodular). As a concrete instance we may take for G the multiplicative group of all matrices

$$\begin{pmatrix} 1 & x & y \\ 0 & 1 & z \\ 0 & 0 & 1 \end{pmatrix}, \quad x,y,z \in \mathbb{R},$$

and for Γ the subgroup of all such matrices with $x,y,z \in \mathbb{Z}$.

(v) In the case of Wiener's example, G is a l.c. abelian group, Γ a closed subgroup such that G/Γ is compact, and N(f) is defined by

$$(4) \qquad N(f) = \sup_{u \in G} \Sigma_{\gamma \in \Gamma} \sup_{x \in K} |f(ux\gamma)|,$$

where K is a fixed compact set such that $G = K.\Gamma$ (Ch. 6, § 2.1). Here the Segal algebra consists of all continuous functions f on G with $N(f) < \infty$, the condition (2) being then automatically satisfied; this follows readily from the equivalence of the norm (4) with the norm $N'(f)$ defined by

$$N'(f) = \Sigma_{\gamma \in \Gamma} \sup_{x \in K} |f(x\gamma)|.$$

In fact, we have

$$N'(f) \leqslant N(f) \leqslant C.N'(f),$$

where C is a constant independent of f (but depending on Γ and K). We may choose C as follows. There are finitely many elements $\gamma_n \in \Gamma$ such that $K^2 \subset \cup_n K.\gamma_n$, since the covering $(K.\gamma)_{\gamma \in \Gamma}$ of G is locally finite; we may take for C the number of these elements γ_n.

(vi) Let G be a l.c. abelian group that is not discrete; let μ be a positive unbounded measure on the dual group $\hat{G}$. For any fixed p, $1 \leqslant p < \infty$, the functions $f \in L^1(G)$ such that $\hat{f} \in L^p(\hat{G},\mu)$ form a Segal algebra, with the norm $\|f\|_S = \|f\|_1 + \|\hat{f}\|_{L^p(\hat{G},\mu)}$. For the case that μ is the Haar measure of $\hat{G}$, this example is familiar, but it is very useful to consider more general measures μ: see (vii) and (viii) below.

(vii) For a l.c. abelian group G, the intersection of all Segal algebras $S^1(G)$ is precisely the set of all (continuous) functions in $L^1(G)$ having a Fourier transform with compact support. The proof is simple. Let $f \in L^1(G)$ be such that Supp $\hat{f}$ is not compact; then there is an infinite sequence $(\hat{a}_n)_{n \geq 1}$ of distinct elements of $\hat{G}$ such that $\hat{f}(\hat{a}_n) \neq 0$ for every n and any compact set in $\hat{G}$ contains at most finitely many $\hat{a}_n$. Let μ be the discrete measure with mass $1/|\hat{f}(\hat{a}_n)|$ at $\hat{a}_n$; then f does not belong to the Segal algebras introduced in (vi), for $p = 1$. On the other hand, the (continuous) functions $f \in L^1(G)$ such that Supp $\hat{f}$ is compact are contained in every Segal algebra $S^1(G)$ (Ch. 6, § 2.2 (iii)).

(viii) As CIGLER [1, p. 276] has shown, a Segal algebra may not admit multiplication by characters; his example is of the type considered in (vi). By an appropriate choice of μ in (vi) it is clear that there are Segal algebras which are not stable under complex conjugation. But the analogous question whether $S^1(G)$, for abelian G, is necessarily stable under the ordinary involution of $L^1(G)$, remains open.

§ 6. Approximate units in Banach algebras: definitions

We introduce here some conditions for a Banach algebra B (with norm $\|.\|$) which generalize well-known properties of $L^1(G)$.

(a) B has approximate left units if, given any $f \in B$ and $\varepsilon > 0$, there is an element $u \in B$ (depending on f and ε) such that

$$\|uf - f\| < \varepsilon$$

(ma) B has multiple approximate left units if, given any finite set $F \subset B$ and $\varepsilon > 0$, there is a $u \in B$ (depending on F and ε) such that

$$\|uf - f\| < \varepsilon \quad \text{for all } f \in F.$$

Condition (ma) may be stated in another way. Suppose B satisfies (ma). Denote by A the set of all pairs $\alpha = (F,n)$ with $F \subset B$ finite and $n = 1,2,\ldots$; A is filtering with respect to the relation '$\leqslant$', where $(F_1,n_1) \leqslant (F_2,n_2)$ means: $F_1 \subset F_2$ and $n_1 \leqslant n_2$. Now for each $\alpha = (F,n)$ in A there is a u_α in B such that

$$\|u_\alpha f - f\| < 1/n \quad \text{for all } f \in F.$$

The family $(u_\alpha)_{\alpha \in A}$, indexed by the filtering set A, has the property:

$$u_\alpha f \to f \quad (\alpha \in A) \text{ for every } f \in B,$$

i.e., given $\varepsilon > 0$, there is an α_ε (depending also on f) such that

$$\|u_\alpha f - f\| < \varepsilon \quad \text{for all } \alpha \text{ with } \alpha_\varepsilon \leqslant \alpha.$$

Conversely, if B contains a family $(u_\alpha)_{\alpha \in A}$, indexed by any filtering set A, possessing this property, then B clearly has multiple approximate units in the sense above.

(b) B has bounded approximate left units if B has approximate left units u (cf. (a)) such that $\|u\| \leq C$, where C is independent of $f \in B$ and $\varepsilon > 0$.

(mb) B has bounded multiple approximate left units if B has multiple approximate left units u (cf. (ma)) such that $\|u\| \leq C$, with C independent of the finite set $F \subset B$ and of $\varepsilon > 0$.

Replacing in (a), (ma), (b), (mb) the word 'left' by 'right' and uf by fu, we obtain the corresponding 'right-hand' conditions (a´), (ma´), (b´), (mb´) for approximate right units, multiple approximate right units,... .

Let us state explicitly the analogous 'two-sided' conditions:

(a") B has approximate two-sided units if, given any $f \in B$ and $\varepsilon > 0$, there is a $u \in B$ such that $\|uf - f\| < \varepsilon$ and $\|fu - f\| < \varepsilon$.

(ma") B has multiple approximate two-sided units if, given any finite set $F \subset B$ and $\varepsilon > 0$, there is a $u \in B$ such that $\|uf - f\| < \varepsilon$ and $\|fu - f\| < \varepsilon$ for all $f \in F$.

It is clear what is to be meant by the conditions (b") and (mb").

It is well-known that $L^1(G)$ has the following property: given finitely many functions $(f_j)_{1\leq j\leq N}$ in $L^1(G)$ and $\varepsilon > 0$, then, if $u \in L^1(G)$ is such that $u \geq 0$, $\int u = 1$ and Supp u is sufficiently small, we have

$$\|u * f_j - f_j\|_1 < \varepsilon \quad \text{and} \quad \|f_j * u - f_j\|_1 < \varepsilon \quad \text{for } 1 \leq j \leq N.$$

Thus the Banach algebra $L^1(G)$ possesses multiple approximate two-sided units, bounded by the constant 1; see, e.g., Ch. 3, § 5.6, where, however, the statement is less explicit.

REMARK 1. In conditions (b), (mb), (b´), (mb´), (b"), (mb"), the word 'bounded' refers to the given norm on B. But we may also consider [multiple] approximate left (right) units bounded with respect to another norm for which B is still a normed algebra: for instance, we may consider the left (right) operator norm on B, defined by

(1) $$f \to \sup_g \|fg\| \ (\sup_g \|gf\|), \ g \in B, \ \|g\| = 1.$$

If B is part of a larger Banach algebra B_1, we may also consider the left (right) operator norm in B_1, restricted to B. It is precisely boundedness with respect to the operator norm which occurs in some applications.

REMARK 2. All definitions apply, of course, also to normed algebras in general. It is readily seen that, if a normed algebra satisfies (b) or (mb), (b´), (mb´), (b"), (mb"), then so does its completion; but the argument breaks down in the case of the corresponding conditions (a), (ma), (a´),... .

We have defined here approximate units of several kinds; among these the bounded multiple ones correspond most closely to the traditional concept which goes back to the paper of Peter-Weyl already mentioned at the end of § 4, where the fundamental importance of approximate units was first stressed, in the case of $\mathcal{K}(G)$, for compact groups G. Approximate units were then discussed, in a systematic way, in A.Weil's book (WEIL [1]; cf. especially pp. 52, 79-80 and 85-86) and have since become a familiar tool. The need for a more general definition first arose in connection with recent work in harmonic analysis concerning Wiener-Ditkin sets (cf. Ch. 7, § 4, especially § 4.10); it will become further apparent in the course of these lectures.

§ 7. Approximate units in Banach algebras: lemmas

The lemmas given here are somewhat technical; they will be required only later (§§ 10, 14, 17).

We consider a Banach algebra B and various kinds of boundedness for approximate units (cf. § 6, Remark 1).

LEMMA 1. If B has approximate left units bounded in the left operator norm, then B also has multiple approximate left units (possibly unbounded).

Proof. We consider the operator A_u:

$$A_u f \;=\; f - uf \quad (u \in B)$$

on B; we have

$$\|A_u f\| \;\leqslant\; (1 + C_u).\|f\|,$$

if C_u is the left operator norm of u on B. By assumption, there is for every $f \in B$ and any $\varepsilon > 0$ an operator A_u ($u = u_{f,\varepsilon}$) such that $\|A_u f\| < \varepsilon$ and with bound $\leqslant 1+C$, where C is independent of f and ε. Now we use induction. Let $N > 1$ elements $f_j \in B$ and $\varepsilon > 0$ be given. By the hypothesis of the induction there is an A_v such that

$$\|A_v f_j\| \;<\; \varepsilon/(1+C) \text{ for } 1 \leqslant j \leqslant N-1;$$

put $f' = A_v f_N$. There is an operator A_w such that $\|A_w f'\| < \varepsilon$ and with bound $\leqslant 1+C$. Thus

$$\|A_w A_v f_j\| \;<\; \varepsilon \quad \text{for all } j, \quad 1 \leqslant j \leqslant N.$$

Moreover, we have

$$A_w A_v \;=\; A_u, \text{ where } u \;=\; w + v - wv,$$

which completes the proof. The induction also shows that for N elements the operator A_u has bound $\leq (1+C)^N$.

This proof is quite analogous to that of Proposition 1 in § 2. Lemma 1 was stated, with a somewhat more restrictive hypothesis, in REITER [2].

LEMMA 2. Let J be a closed two-sided ideal of B. If B/J (provided with the quotient norm) has [multiple] approximate left units and if J has [multiple] approximate left units bounded in the <u>left operator norm of the algebra</u> B, then B has [multiple] approximate left units.

<u>Proof</u>. We shall write the argument for the case of multiple approximate left units; the case of 'simple' approximate left units corresponds to N = 1 in the following. Let elements f_n, $1 \leq n \leq N$, of B and $\varepsilon > 0$ be given. Denote by T the canonical homomorphism $B \to B/J$ and consider $Tf_n \in B/J$: by hypothesis there is an element $v' \in B/J$ such that

(1) $$\|v'.Tf_n - Tf_n\|_{B/J} < \varepsilon \qquad 1 \leq n \leq N.$$

Take $v \in B$ such that

(2) $$v' = Tv$$

and put

(3) $$g_n = f_n - v.f_n \qquad 1 \leq n \leq N.$$

There are elements $j_n \in J$ such that

(4) $$\|g_n - j_n\| < \|Tg_n\|_{B/J} + \varepsilon \qquad 1 \leq n \leq N.$$

By hypothesis, there is an element $w \in J$ such that

(5) $$\|j_n - w.j_n\| < \varepsilon \qquad 1 \leq n \leq N$$

and w is bounded in the left operator norm of B:

(6) $$\|wz\| \leqslant C.\|z\| \qquad \text{for all } z \in B,$$

where C is independent of $\{j_1,\ldots,j_n\}$ and ε. Now we write

$$g_n - w.g_n = (g_n - j_n) + (j_n - w.j_n) + w.(j_n - g_n),$$

whence by (4), (5), (6)

$$\|g_n - w.g_n\| \leqslant \|Tg_n\|_{B/J} + \varepsilon + \varepsilon + C.(\|Tg_n\|_{B/J} + \varepsilon),$$

or, since $\|Tg_n\|_{B/J} < \varepsilon$ (cf. (1), (2), (3)),

$$\|g_n - w.g_n\| < (3 + 2C).\varepsilon.$$

Writing here again the original expression (2) for g_n, we obtain

(7) $$\|f_n - u.f_n\| < (3 + 2C).\varepsilon, \qquad 1 \leqslant n \leqslant N,$$

where

(8) $$u = w + v - wv.$$

This completes the proof, the coefficient of ε in (7) being independent of ε.

LEMMA 3. Let J be a closed two-sided ideal of B. If B/J and J have <u>bounded</u> [multiple] approximate left units, then B has bounded [multiple] approximate left units.

<u>Proof</u>. This is entirely like that of lemma 2 above, with only minor changes and additions; we consider again the 'multiple' case (the 'simple' case corresponding to N = 1). We can, by hypothesis, choose the element $v' \in B/J$ figuring in (1) such that it satisfies the additional condition.

(9) $$\|v'\|_{B/J} \leqslant C',$$

where C' <u>is independent of</u> (Tf_n, and hence of) <u>the elements</u> f_n,

$1 \leqslant n \leqslant N$, <u>and of</u> ε. Then we can choose in (2) the element $v \in B$ such that

$$\|v\| \leqslant C'+1,$$

say. The proof then proceeds along the same lines as before; moreover, (6) is then replaced by

$$\|w\| \leqslant C,$$

where C <u>is independent of</u> f_n, $1 \leqslant n \leqslant N$, <u>and of</u> ε. For the element $u \in B$ appearing in (7) and (8) we then have

$$\|u\| \leqslant C + (C'+1) + C.(C'+1),$$

which completes the proof.

REMARK 1. All three lemmas have a corresponding 'right-hand' version; the proofs are 'symmetric'.

REMARK 2. The three lemmas remain valid, of course, for normed algebras; see also § 6, Remark 2.

§ 8. Approximate units in Segal algebras

Segal algebras possess approximate units with some special properties, of importance in the applications.

PROPOSITION 1.

(i) Every Segal algebra has multiple approximate left units u that have L^1-norm 1 and are self-adjoint, i.e. satisfy

$$u^* = u. \tag{1}$$

(ii) A symmetric Segal algebra has multiple approximate two-sided units that have L^1-norm 1 and are self-adjoint.

(iii) A pseudosymmetric Segal algebra has multiple approximate two-sided units that are positive, of L^1-norm 1, and self-adjoint.

The fact that we can choose the approximate units self-adjoint will be essential later (cf. § 12, Remark 2).

Proof.

(i) Let a finite set $F \subset S^1(G)$ and $\varepsilon > 0$ be given. There is a nd. U of e in G such that

$$\|L_y f - f\|_S < \varepsilon/2 \quad \text{for all } f \in F \text{ and all } y \in U.$$

Let $u_1 \in \mathcal{K}(G)$ be such that $u_1 \geq 0$, $\int u_1 = 1$ and Supp $u_1 \subset U$; then (cf. § 4, Lemma 1, (i))

$$u_1 * f - f = \int u_1(y).(L_y f - f)dy,$$

whence

$$\|u_1 * f - f\|_S \leq \int u_1(y).\|L_y f - f\|_S dy < \varepsilon/2 \qquad f \in F. \tag{2}$$

Now choose $u \in S^1(G)$ such that

$$\|u - u_1\|_1 < \varepsilon/(2M), \tag{3}$$

where $M = \max \|f\|_S$ $(f \in F)$; such a $u \in S^1(G)$ exists, since $S^1(G)$ is dense in $L^1(G)$. Then we have

$$\|u * f - f\|_S \leqslant \|u - u_1\|_1 . \|f\|_S + \|u_1 * f - f\|_S$$

and hence

$$(4) \qquad \|u * f - f\|_S \leqslant \varepsilon/2 + \varepsilon/2 = \varepsilon \qquad f \in F.$$

Clearly we can choose $u \in S^1(G)$ such that in addition to (3) we also have $\|u\|_1 = 1$ (L^1-norm!). Thus it is quite easy to prove part (i) without the condition (1); to obtain (1), more work is needed.

First we make a special choice for the function $u_1 \in \mathcal{K}(G)$ considered above: we take

$$(5) \qquad u_1 = v_1^* * v_1,$$

where $v_1 \in \mathcal{K}(G)$, $v_1 \geqslant 0$, $\int v_1 = 1$ and Supp $v_1 \subset V$, the nd. V of $e \in G$ being chosen such that $V^{-1}.V \subset U$, with U as before; then u_1 has the properties stated above and (2) holds.

Next we can choose, given any $\varepsilon' > 0$, a function $v \in S^1(G)$ such that

$$(6) \qquad \|v - v_1\|_1 < \varepsilon'.$$

For our purposes we take

$$(7) \qquad \varepsilon' = \min(1, \varepsilon/(12M)),$$

with the same M as in (3); the reason will appear in a moment.

Then we put

$$(8) \qquad u = v^* * v / \|v^* * v\|_1.$$

The function u is in $S^1(G)$, since $S^1(G)$ is a left ideal of $L^1(G)$, it has L^1-norm 1, and is self-adjoint. We shall now show that u satisfies (3), if ε' is chosen as in (7); then u will also satisfy (4) and the proof will be complete.

To obtain (3), we write

$$u - u_1 = v^* * v - v_1^* * v_1 + [(1/\|v^* * v\|_1) - 1].v^* * v,$$

whence

$$\|u - u_1\|_1 \leqslant \|v^* * v - v_1^* * v_1\|_1 + |1 - \|v^* * v\|_1|$$

or, since $1 = \int v_1^* * v_1 = \|v_1^* * v_1\|_1$,

(9) $$\|u - u_1\|_1 \leqslant 2.\|v^* * v - v_1^* * v_1\|_1.$$

Now

$$v^* * v - v_1^* * v_1 = v^* * (v - v_1) + (v - v_1)^* * v_1,$$

thus by (6)

$$\begin{aligned}\|v^* * v - v_1^* * v_1\|_1 &\leqslant (1 + \varepsilon').\varepsilon' + \varepsilon'.1 \\ &\leqslant 3\varepsilon' \quad \text{if} \quad \varepsilon' \leqslant 1.\end{aligned}$$

Hence (9) yields

(10) $$\|u - u_1\|_1 \leqslant 6\varepsilon' \ (\text{if } \varepsilon' \leqslant 1).$$

Choosing ε' as in (7), we obtain (3), and hence (4), so that the proof of part (i) is complete.

(ii) Let again a finite set $F \subset S^1(G)$ and $\varepsilon > 0$ be given. There is a nd. U of $e \in G$ such that

(11) $\|L_y f - f\|_S < \varepsilon/2$ <u>and</u> $\|R_y f - f\|_S < \varepsilon/2$ for all $f \in F$ and all $y \in U$.

Then for $u_1 \in \mathcal{K}(G)$ such that $u_1 \geqslant 0$, $\int u_1 = 1$ and Supp $u_1 \subset U$ we have by (2) and the principle of symmetry

$$\|u_1 * f - f\|_S < \varepsilon/2 \quad \underline{\text{and}} \quad \|f * u_1 - f\|_S < \varepsilon/2 \qquad f \in F.$$

Now we choose u_1, v and u as in (i) (cf. (8)), so that (3) holds;

then we obtain a stronger form of (4), namely

(12) $\|u*f - f\|_S < \varepsilon$ <u>and</u> $\|f*u - f\|_S < \varepsilon \quad f \in F$,

by using also the principle of symmetry. Thus part (ii) is proved.

(iii) If $S^1(G)$ is pseudosymmetric, we can use the condition (S^o) of § 4; moreover, we may assume that the functions $u \in S^1(G)$ mentioned in (S^o) also satisfy $u^* = u$ (besides $u \geq 0$, $\int u = 1$): for otherwise we may replace u by $u^* * u$, which belongs to $S^1(G)$ if u does, and also satisfies $u^* * u \geq 0$, $\int u^* * u = 1$; also Supp $u^* * u$ is arbitrarily small if Supp u is small enough.

Choosing U so that (11) holds, and then $u \in S^1(G)$ such that $u \geq 0$, $\int u = 1$, $u^* = u$ and Supp $u \subset U$, we at once obtain (12) (using also § 4, Lemma 1, (ii)). Thus for the pseudosymmetric case the proof is rather short.

COROLLARY. (i) Every <u>closed</u> left ideal of a Segal algebra $S^1(G)$ is a left ideal in $L^1(G)$; (ii) if $S^1(G)$ is symmetric, then every <u>closed</u> right ideal of $S^1(G)$ is a right ideal in $L^1(G)$.

<u>Proof</u>. (i) Consider $f \in I$, a closed left ideal of $S^1(G)$, and $g \in L^1(G)$. For any $\varepsilon > 0$ there is a $u \in S^1(G)$ such that

$$\|u*f - f\|_S < \varepsilon.$$

Then $g*u$ is in $S^1(G)$, hence $(g*u)*f$ is in I. Now

$$\|g*u*f - g*f\|_S \leq \|g\|_1 . \varepsilon,$$

thus $g*f$ is in I, since I is closed. (ii) The proof is symmetrical.

In connection with Proposition 1 the case of a l.c. <u>abelian</u> group merits special consideration.

PROPOSITION 2. If G is a l.c. abelian group, then every Segal algebra on G has multiple approximate units of L^1-norm 1 and having a <u>positive</u> Fourier transform with <u>compact</u> support.

<u>Proof.</u> This is the same as the proof of part (i) of Proposition 1: we only observe that in (6) we may take v such that $\hat{v}$ has compact support; this is possible since such functions v are dense in $L^1(G)$ (Ch. 5, § 4.1) and belong to $S^1(G)$ (Ch. 6, § 2.2 (iii)). The function $u \in S^1(G)$ defined by (8) and which satisfies (4) then has all the required properties.

The approximate units of Proposition 2 are positive-definite functions and hence, in particular, self-adjoint. Thus in the abelian case Proposition 1 is contained in Proposition 2. For the case $S^1(G) = L^1(G)$, Proposition 2 is more or less classical.

In the case of Segal algebras, the proof of some facts that are quite obvious for $L^1(G)$ require more subtle methods, as we have just seen. To what extend other familiar properties of $L^1(G)$ extend to Segal algebras is a matter that calls for further research, even in the case of abelian groups (cf. § 5, (viii)).

§ 9. Some general results on Segal algebras

First we establish the analogue of a familiar property of $L^1(G)$.

PROPOSITION 1. In a Segal algebra $S^1(G)$ the closed left ideals coincide with the closed left invariant linear subspaces; if $S^1(G)$ is symmetric, then the closed right ideals coincide with the closed right invariant linear subspaces.

Proof. Let I be a closed left invariant linear subspace of $S^1(G)$. Consider $f \in I$, $h \in S^1(G)$: we have $h * f = \int h(y)L_y f\, dy$ (§ 4, Lemma 1, (i)); moreover,

$$\|\int h(y)L_y f\, dy - \Sigma_n a_n L_{y_n} f\|_S < \varepsilon$$

for properly chosen complex coefficients a_n and elements $y_n \in G$ (Ch. 3, § 5.9). Hence $h * f$ lies in I, since I is closed. Conversely, let I be a closed left ideal of $S^1(G)$. Using the existence of approximate left units in $S^1(G)$ (§ 8, Proposition 1, (i)), we can prove in exactly the same way as in Ch. 3, § 5.7 (i) that I is invariant under left translations. If $S^1(G)$ is symmetric, the proof for right ideals goes by symmetry.

Next we consider the relation between the closed right ideals of $S^1(G)$ and the closed right ideals of $L^1(G)$, and correspondingly for closed left ideals: we prove an extension to general l.c. groups of a result given in Ch. 6, § 2.4 for the case of l.c. abelian groups.

THEOREM 1. Let $S^1(G)$ be any symmetric Segal algebra. Then every closed right [left] ideal I_S of $S^1(G)$ is of the form $I \cap S^1(G)$, where I is a unique closed right [left] ideal of $L^1(G)$; I is the closure of I_S in $L^1(G)$. If $S^1(G)$ is pseudosymmetric, the assertion remains true for closed right ideals.

Proof.

(i) First we show the following. If $S^1(G)$ is symmetric or pseudosymmetric, then for any $f \in L^1(G)$ and $\varepsilon > 0$ there is a u_S in $S^1(G)$ (not merely in $L^1(G)$!) such that

(1) $$\|f * u_S - f\|_1 < \varepsilon.$$

Indeed, since $S^1(G)$ is dense in $L^1(G)$, there is an $f_S \in S^1(G)$ satisfying

$$\|f - f_S\|_1 < \varepsilon/3.$$

Next there is a $u_S \in S^1(G)$ such that

$$\|f_S * u_S - f_S\|_S < \varepsilon/3$$

and having L^1-norm 1 (§ 8, Proposition 1, (ii) and (iii)); then we have a fortiori

$$\|f_S * u_S - f_S\|_1 < \varepsilon/3.$$

Hence we obtain

$$\begin{aligned}\|f * u_S - f\|_1 &\leqslant \|f - f_S\|_1 . \|u_S\|_1 + \|f_S * u_S - f_S\|_1 + \|f_S - f\|_1 \\ &< (\varepsilon/3).1 + \varepsilon/3 + \varepsilon/3 = \varepsilon.\end{aligned}$$

(ii) We can now prove part of the assertion of the theorem. Let $S^1(G)$ be symmetric or pseudosymmetric. Let I be a closed right ideal of $L^1(G)$ and put $I_S = I \cap S^1(G)$. Then I_S is a closed right ideal of $S^1(G)$ and I is the closure of I_S in $L^1(G)$.

That I_S is a closed right ideal of $S^1(G)$, is clear from the fact that the Segal norm majorizes the L^1-norm. Now let any $f \in I$ and $\varepsilon > 0$ be given; then we can choose $u_S \in S^1(G)$ such that (1) holds. The convolution $f * u_S$ belongs to I, since I is a right ideal, and to $S^1(G)$, since $S^1(G)$ is a left ideal in $L^1(G)$; thus $f * u_S$ is in I_S, and $f \in I$ lies in the L^1-closure of I_S. Hence I, being closed, coincides with this closure.

(iii) We now prove the converse of (ii). Let $S^1(G)$ be symmetric or pseudosymmetric and let I_S be a closed right ideal of $S^1(G)$; let I be the closure of I_S in $L^1(G)$. Then I is a closed right ideal of $L^1(G)$ and the equality $I_S = I \cap S^1(G)$ holds.

That I is a closed <u>right ideal of</u> $L^1(G)$ follows from the fact that $S^1(G)$ is dense in $L^1(G)$. The following elegant reasoning, due to BURNHAM [1,2], shows that $I_S = I \cap S^1(G)$. Obviously

$$I_S \subset I \cap S^1(G).$$

To establish the opposite inclusion, consider any $f \in I \cap S^1(G)$. In order to prove that $f \in I_S$, it suffices to show that any $S^1(G)$-neighbourhood of f contains points of I_S (since I_S is closed in $S^1(G)$, by assumption). Now, given $\varepsilon > 0$, there is a $u \in S^1(G)$ such that

$$\|f * u - f\|_S < \varepsilon, \tag{2}$$

since f is in $S^1(G)$ and $S^1(G)$ has approximate right units (§ 8, Proposition 1, (ii) and (iii)). Since f is also in I and I is the L^1-closure of I_S by assumption, there is a $g \in I_S$ such that

$$\|f - g\|_1 < \varepsilon/\|u\|_S.$$

This implies by a general property of Segal algebras

$$\|f * u - g * u\|_S < \varepsilon.$$

Combining this with (2), we get

$$\|g * u - f\|_S < 2\varepsilon.$$

Here $g * u$ is in I_S, as I_S is by assumption a right ideal of $S^1(G)$. Since $\varepsilon > 0$ was arbitrary, the proof of (iii) is complete.

(iv) Finally we observe that, if $S^1(G)$ is symmetric, the assertion of the theorem for closed <u>left</u> ideals results from the proof above by symmetry. This concludes the proof of Theorem 1.

Theorem 1 is, essentially, a particular case of a result due to BURNHAM [1,2]; his work will also be found (it is hoped) in his future publications. We only mention here another of BURNHAM's results, specialized to Segal algebras: a Segal algebra $S^1(G)$ cannot have approximate right units that are bounded in the Segal norm unless $S^1(G) = L^1(G)$; if $S^1(G)$ is symmetric, the same holds for approximate left units (but here the proof depends in an essential way on the symmetry!). We shall not need these results, however, and leave them as exercises for the interested reader.

We have now established some basic properties of Segal algebras - in particular of the symmetric and pseudosymmetric ones - and have reached a point where we can extend the approach begun in § 1 for $L^1(G)$ to Segal algebras, and can prove results not only for $L^1(G)$ but also for Segal algebras. This will be the task of the following sections.

§ 10. Certain ideals in Segal algebras

In § 1 we defined, for any closed subgroup $H \subset G$, two linear subspaces $D_H L^1(G)$ and $D'_H L^1(G)$ of $L^1(G)$; let us now consider their analogues for Segal algebras.

DEFINITION. Let $S^1(G)$ be any Segal algebra and let H be a closed subgroup of G. Then we define $D_H S^1(G)$ as the closed linear subspace generated in $S^1(G)$ by the set

$$\{f - L_y f \mid f \in S^1(G),\ y \in H\}.$$

If $S^1(G)$ is right invariant - thus in particular if $S^1(G)$ is symmetric or pseudosymmetric - we can also introduce $D'_H S^1(G)$, the closed linear subspace generated in $S^1(G)$ by the set

$$\{f - R_y f \mid f \in S^1(G),\ y \in H\}.$$

We remark that, in the definition above, 'closed' refers, of course, to the topology of $S^1(G)$ induced by the Segal norm.

$D_H S^1(G)$ <u>is a</u> (closed) <u>right ideal of</u> $S^1(G)$ and, if $S^1(G)$ is right invariant, $D'_H S^1(G)$ <u>is a</u> (closed) <u>left ideal</u>; this is shown just as in the L^1-case (cf. § 1, (5)). It also follows directly from the definition that, if $S^1(G)$ is symmetric, then $D_H S^1(G)$ is a (closed) right invariant subspace, and that $D'_H S^1(G)$ is a (closed) left invariant subspace for right invariant $S^1(G)$.

The relations between $D_H S^1(G)$ and $D'_H S^1(G)$ will be discussed in §§ 11, 12; in the case of $S^1(G)$ these relations are more subtle than for $L^1(G)$ (§ 1).

In this section we prove the following two results.

THEOREM 1. Let $S^1(G)$ be a Segal algebra and let H be a closed subgroup of G having the property P_1. Then $D_H S^1(G)$ has multiple approximate <u>left</u> units, bounded in L^1-norm by the constant 2. If $S^1(G)$ is symmetric, then $D'_H S^1(G)$ has multiple approximate <u>right</u> units, bounded in L^1-norm by 2.

THEOREM 2. Let $S^1(G)$ be a Segal algebra and let H be <u>any</u> closed subgroup of G. If $D_H S^1(G)$ has approximate <u>right</u> units bounded in the right operator norm of $D_H S^1(G)$ (cf. § 6, (1)), then H has the property P_1. If $S^1(G)$ is right invariant and $D'_H S^1(G)$ has approximate <u>left</u> units bounded in the left operator norm of $D'_H S^1(G)$, then the same conclusion holds.

REMARK 1. It should be noted that in Theorem 2 the approximate units are not assumed to be multiple ones, and that boundedness in the right [left] operator norm of $D_H S^1(G)$ [$D'_H S^1(G)$] is a less restrictive condition than boundedness in the right [left] operator norm of $S^1(G)$ and, if $S^1(G)$ is symmetric, boundedness in the L^1-norm [for the left operator norm, this holds of course for general $S^1(G)$]. Further it will be observed that, in the strict sense, Theorem 2 is not a converse of Theorem 1, as 'right' and 'left' are interchanged; but under certain conditions this becomes true (cf. § 12, (iii)-(v)).

<u>Proof of Theorem 1</u>. This depends on the following fact. Let $S^1(G)$ be a Segal algebra and let H be a closed P_1-subgroup of G. Then, given finitely many functions in $D_H S^1(G)$, say $(f_j)_{1\leq j\leq N}$, and $\varepsilon > 0$, there is an operator L of the form

(1) $$L = \Sigma_n c_n L_{\xi_n},\ c_n > 0,\ \Sigma_n c_n = 1,\ \xi_n \in H,$$

such that

(2) $$\|Lf_j\|_S < \varepsilon \qquad \text{for } j = 1,\ldots,N.$$

Moreover, if $S^1(G)$ is symmetric, then there is also an operator

(3) $$R = \Sigma_n c_n R_{\xi_n},\ c_n > 0,\ \Sigma_n c_n = 1,\ \xi_n \in H$$

such that

(4) $$\|Rf_j\|_S < \varepsilon \qquad \text{for } j = 1,\dots,N.$$

(The coefficients c_n and the elements ξ_n in (1) and (3) are, of course, not necessarily the same).

REMARK 2. For the case of $L^1(G)$ the relation (2) has already been proved by the methods of classical analysis (cf. § 2, especially Proposition 1); moreover, (4) will not be needed in the case of $L^1(G)$. The proof of (2) and (4) for Segal algebras will be given below (Proposition 1).

Let us now show how Theorem 1 follows from (2) and (4); the reader may consider the case $S^1(G) = L^1(G)$, for concreteness, in view of Remark 2 above.

Consider first $D_H S^1(G)$. Let $(f_j)_{1 \leq j \leq N}$ in $D_H S^1(G)$ and $\varepsilon > 0$ be given. There is a $u \in S^1(G)$ such that

(5) $$\|u * f_j - f_j\|_S < \varepsilon \qquad j = 1,\dots,N$$

and

(6) $$\|u\|_1 = 1,$$

by § 8, Proposition 1, (i) (here the fact that u may also be chosen self-adjoint is not needed). Now, choosing L according to (1) and (2), we have in view of (5), since L is a contraction,

(7) $$\|L(u * f_j)\|_S < 2\varepsilon \qquad j = 1,\dots,N.$$

Next we observe that

$$(u - Lu) * f_j - f_j = u * f_j - f_j - L(u * f_j).$$

Hence, putting

$$(8) \qquad v = u - Lu,$$

we obtain from (5) and (7)

$$(9) \qquad \|v * f_j - f_j\|_S < 3\varepsilon \qquad j = 1,\ldots,N.$$

Moreover, v belongs to $D_H S^1(G)$, for we may write (cf. (1))

$$v = \Sigma_n c_n (u - L_{\xi_n} u)$$

and $u - L_\xi u$ lies in $D_H S^1(G)$ for $u \in S^1(G)$, $\xi \in H$, by definition. Also, for the L^1-norm of v we have, in view of (8) and (6),

$$(10) \qquad \|v\|_1 \leqslant \|u\|_1 + \|Lu\|_1 \leqslant 2.$$

(9) and (10) show that $D_H S^1(G)$ has multiple approximate <u>left</u> units of L^1-norm $\leqslant 2$.

In the case $S^1(G) = L^1(G)$ we then obtain at once that $D_H' L^1(G)$ has multiple approximate <u>right</u> units of L^1-norm $\leqslant 2$: this follows simply from § 1, (7) and familiar properties of the involution in $L^1(G)$.

In the general case, when $S^1(G)$ is a symmetric Segal algebra, we obtain the existence of multiple approximate right units in $D_H' S^1(G)$, with L^1-norm $\leqslant 2$, from relation (4) in a way entirely analogous to that above, by applying the principle of symmetry.

It remains to prove the relations (2) and (4). This is done by means of Glicksberg's theorem (Ch. 8, § 6.1), which yields the following general result.

PROPOSITION 1. Let H be a l.c. P_1-group. Suppose H acts on the Banach space B (with norm $\|.\|$) by means of linear operators A_ξ, $\xi \in H$, such that

$$(a) \quad A_{\xi_1 \xi_2} = A_{\xi_1} A_{\xi_2}; \quad (b) \quad \|A_\xi f\| \leqslant \|f\| \ (f \in B);$$

(c) the mapping $\xi \to A_\xi f$ of H into B is continuous, for each $f \in B$.

Let $D_H B$ be the closed linear subspace of B generated by the set

$$\{f - A_\xi f \mid f \in B,\ \xi \in H\}.$$

Then, given finitely many functions $f_j \in D_H B$, $1 \leqslant j \leqslant N$, and $\varepsilon > 0$, there is an operator A of the form

(11) $$A = \Sigma_n c_n A_{\xi_n},\ c_n > 0,\ \Sigma_n c_n = 1,\ \xi_n \in H,$$

such that

(12) $$\|Af_j\| < \varepsilon \qquad \text{for } j = 1,\ldots,N.$$

Proof. For N = 1 this is part of Glicksberg's theorem (cf. loc. cit.), since P_1 implies (M). Then we obtain the result for arbitrary N by induction: the method is precisely the same as that used for proving Proposition 1 in § 2. But it is also possible to get around the induction by applying Glicksberg's theorem directly to the product B^N, which is a Banach space on which H acts in an obvious way.

Proposition 1 now yields both (2) and (4) if we put $B = S^1(G)$, $A_\xi = L_\xi$ or $A_\xi = R_{\xi^{-1}}$, respectively: note that the conditions (a), (b), (c) for the operators A_ξ are then satisfied - in the case of $R_{\xi^{-1}}$ by virtue of the assumption that $S^1(G)$ is symmetric (note also that in (3) we used R_ξ, not $R_{\xi^{-1}}$, but this clearly makes no difference).

Thus we have proved (2) and (4) in the general case of Segal algebras, which completes the proof of Theorem 1.

REMARK 3. For symmetric or pseudosymmetric Segal algebras the statements (1), (2) can be deduced from their validity for the special case $S^1(G) = L^1(G)$, and can thus be proved without recourse to Glicksberg's theorem (cf. Remark 2); likewise for the statements (3), (4) in the case of symmetric Segal algebras.

<u>Proof</u>. Let $(f_j)_{1 \leq j \leq N}$ in $D_H S^1(G)$ and $\varepsilon > 0$ be given. As $S^1(G)$ is symmetric or pseudosymmetric, there is a $v \in S^1(G)$ such that

(13) $$\|f_j * v - f_j\|_S < \varepsilon/2 \qquad j = 1,\ldots,N$$

(cf. § 8, Proposition 1, parts (ii) and (iii)). By § 2, Proposition 1, there is an $L \in \mathcal{L}_H$ such that

$$\|Lf_j\|_1 < \varepsilon/(2\|v\|_S) \qquad j = 1,\ldots,N,$$

since clearly

(14) $$D_H S^1(G) \subset D_H L^1(G).$$

Then we have, by a general property of Segal algebras,

$$\|(Lf_j) * v\|_S < \varepsilon/2$$

or, since $(Lf_j) * v = L(f_j * v)$,

$$\|L(f_j * v)\|_S < \varepsilon/2.$$

Since L is a contraction on $S^1(G)$, we have from (13)

$$\|L(f_j * v) - Lf_j\|_S < \varepsilon/2.$$

Combining the last two inequalities, we obtain

$$\|Lf_j\|_S < \varepsilon \qquad j = 1,\ldots,N,$$

which proves (1), (2), for symmetric or pseudosymmetric $S^1(G)$ (H being a P_1-subgroup).

We can obtain (3), (4), in the case of a symmetric Segal algebra, by an entirely 'symmetric' proof.

REMARK 4. It follows from Remark 3 that <u>for symmetric or pseudo-symmetric Segal algebras the assertion of Theorem</u> 1 <u>can be established without recourse to Glicksberg's theorem</u>. Whether it is possible to deduce (1) and (2) for arbitrary Segal algebras from their validity for $L^1(G)$ remains an open question (in the case of (3) and (4) this question does not arise!); this is of interest for the proof of Theorem 1, as we have seen.

Proof of Theorem 2. Consider the case of $D_H S^1(G)$. From the assumption that $D_H S^1(G)$ has approximate right units bounded in the right operator norm it follows that $D_H S^1(G)$ has multiple approximate right units (possibly unbounded): see § 7, Lemma 1, and Remark 1 at the end of § 7). Let any finite subset E of the subgroup H, and $\varepsilon > 0$, be given. Choose any $f \in S^1(G)$ such that

$$(15) \qquad \int f(x)dx = 1$$

and consider the functions $L_y f - f$, $y \in E$, which lie in $D_H S^1(G)$. By hypothesis, there is a $u \in D_H S^1(G)$ such that

$$\|(L_y f - f) * u - (L_y f - f)\|_S < \varepsilon \quad \text{for each } y \in E.$$

Putting

$$(16) \qquad g = f - f * u,$$

we can write

$$L_y g - g = L_y f - f - (L_y f - f) * u,$$

thus $\|L_y g - g\|_S < \varepsilon$ for each $y \in E$. Hence we have a fortiori for the L^1-norm

$$(17) \qquad \|L_y g - g\|_1 < \varepsilon \qquad y \in E.$$

We also have

$$(18) \qquad \int g(x)dx = 1,$$

by (15), (16) and the fact that $\int u = 0$ (because $u \in D_H S^1(G)$). Now put

$$s(x) = |g(x)|/\|g\|_1.$$

Then

$$s \in L^1(G),\ s \geq 0,\ \int s = 1,$$

and (17) yields, since $\|g\|_1 \geq 1$ by (18),

$$\|L_y s - s\|_1 < \varepsilon \qquad y \in E.$$

We have thus shown that the property $P_*(G,H)$ of § 3 holds, and an application of § 3, Proposition 1, completes the proof for the case of $D_H S^1(G)$. For $D_H' S^1(G)$ the proof is entirely 'symmetrical'. Thus the proof of Theorem 2 is complete.

Theorem 1 and 2 generalize results for $L^1(G)$ (REITER [2]); for Theorem 1 see also Remark 4 above, and for Theorem 2 see Remark 1 in respect of the boundedness condition for the approximate units.

Let us emphasize that part of Theorem 1, and Theorem 2, hold true for <u>general</u> Segal algebras. In the developments to be given next, we shall have to restrict ourselves to symmetric or pseudosymmetric Segal algebras.

REMARK 5. In this section, in Remark 3, a phenomenon of structural significance has appeared for the first time: sometimes a result for Segal algebras may first be proved for $L^1(G)$ and can then be deduced for Segal algebras - but sometimes such a ' reduction to $L^1(G)$ ' is not feasible and then a direct approach has to be devised; again, sometimes both methods may be applicable.

§ 11. Canonical representation of the ideals

Let us now consider the relations between the ideals $D_H S^1(G)$ and $D_H' S^1(G)$, introduced in § 10, and their prototypes $D_H L^1(G)$ and $D_H' L^1(G)$ defined for $L^1(G)$ in § 1.

PROPOSITION 1. If $S^1(G)$ is symmetric, then the <u>canonical representations</u>

(1) $$D_H S^1(G) = D_H L^1(G) \cap S^1(G),$$

(2) $$D_H' S^1(G) = D_H' L^1(G) \cap S^1(G)$$

hold. Moreover, (1) also holds if $S^1(G)$ is pseudosymmetric.

<u>Proof</u>. Since $S^1(G)$ is dense in $L^1(G)$, it is clear that the closure of $D_H S^1(G)$ in $L^1(G)$ is just $D_H L^1(G)$, for <u>any</u> $S^1(G)$; likewise the closure of $D_H' S^1(G)$ in $L^1(G)$ is simply $D_H' L^1(G)$. Combining this with Theorem 1 of § 9, we see (recalling also § 8, Proposition 1, (ii), (iii)) that (1) and (2) hold under the conditions stated above.

<u>COROLLARY of Proposition</u> 1. If $S^1(G)$ is symmetric and stable under involution, then the following relation holds:

(3) $$D_H' S^1(G) = [D_H S^1(G)]^*.$$

<u>Proof</u>. This follows at once from Proposition 1, combined with § 1, (7), and from the assumed stability of $S^1(G)$ under involution.

REMARK 1. If $S^1(G)$ is star-symmetric, then of course (3) can be proved directly, i.e. without the use of the representation formulae (1) and (2) above, in exactly the same way as the relation § 1, (7). But it is not known whether for a Segal algebra symmetry and stability

under involution imply (and hence are equivalent to) star-symmetry.

PROPOSITION 2. If $S^1(G)$ is symmetric, then the relations

(4) $$D_H S^1(G) = [J^1(G,H)]^* \cap S^1(G)$$

(5) $$D'_H S^1(G) = J^1(G,H) \cap S^1(G)$$

hold, where $J^1(G,H)$ is defined by § 1, (29). Moreover (4) also holds if $S^1(G)$ is pseudosymmetric.

Proof. In view of the relations § 1, (34) and (33), Proposition 2 is equivalent to Proposition 1.

It will now be shown that Proposition 2 can also be established directly, i.e. without proving first Proposition 1 by means of Theorem 1 of § 9, and then appealing to the special relations § 1, (34) and (33), for $L^1(G)$: we shall, in effect, give a proof for Proposition 2 that reduces to the proof of those relations when $S^1(G) = L^1(G)$. This direct proof is by the method of linear functionals.

We denote the continuous linear functionals on $S^1(G)$, i.e. the elements of the dual space $S^1(G)'$, by ϕ_S; we write

(6) $$\langle f,\phi_S\rangle_S$$

for the value of ϕ_S at the 'point' $f \in S^1(G)$. In the case of a bounded, measurable function ϕ on G, we write, by abuse of notation,

(7) $$\langle f,\phi\rangle_S = \int f(x)\overline{\phi(x)}dx \qquad f \in S^1(G).$$

In the case $S^1(G) = L^1(G)$ the subscript S in (7) is simply omitted, whereby (7) reduces to § 1, (12).

We first prove a lemma that will also be useful in another context (§ 16, Lemma 1).

LEMMA 1. For every Segal algebra $S^1(G)$ and any $\phi_S \in S^1(G)'$ the relation

$$(8)\qquad \langle f * g, \phi_S\rangle_S = \int f(y)\langle L_y g, \phi_S\rangle_S \, dy \qquad f \in L^1(G),\ g \in S^1(G)$$

holds. If $S^1(G)$ is symmetric, then the relation

$$(9)\qquad \langle g * f, \phi_S\rangle_S = \int f(y)\langle R_y g, \phi_S\rangle_S \, dy \qquad f \in L^1(G),\ g \in S^1(G)$$

also holds.

Proof. This is a simple application of § 4, Lemma 1, parts (i) and (iii) (with a change in notation), and of a fundamental relation for vector integrals (§ 4, (12)): we have thereby

$$\langle \int f(y)L_y g\,dy, \phi_S\rangle_S = \int \langle f(y)L_y g, \phi_S\rangle_S \, dy,$$

$$\langle \int f(y)R_y g\,dy, \phi_S\rangle_S = \int \langle f(y)R_y g, \phi_S\rangle_S \, dy,$$

and the terms on the right give the corresponding terms in (8) and (9).

We can now give the second proof of Proposition 2.

To prove (4), we first establish the inclusion

$$(10)\qquad D_H S^1(G) \subset [J^1(G,H)]^* \cap S^1(G).$$

For this, we show: for any $f \in S^1(G)$ we have

$$(11)\qquad L_y f - f \in [J^1(G,H)]^* \cap S^1(G) \qquad y \in H.$$

That $L_y f - f$ lies in $S^1(G)$ is trivial. To show that $L_y f - f$ lies in $[J^1(G,H)]^*$, for $y \in H$, we have to verify that (cf. § 1, (29))

$$(12)\qquad T_{H,q}((L_y f - f)^*) = 0 \text{ if } y \in H.$$

Since $(L_y f - f)^* = R_{y^{-1}}(f^*) - f^*$ (cf. § 1, (9)), we see by § 1, (30) that (12) is true. Thus (11) holds; as the right-hand side of (11) is clearly a closed linear subspace of $S^1(G)$, this yields (10).

To prove the inclusion opposite to (10), consider any functional

$\phi_S \in S^1(G)'$ that vanishes on $D_H S^1(G)$: thus ϕ_S satisfies for all $g \in S^1(G)$

$$(13) \qquad \langle L_y g, \phi_S \rangle_S = \langle g, \phi_S \rangle_S \qquad y \in H.$$

Let us put, for fixed $g \in S^1(G)$,

$$(14) \qquad \psi(x) = \langle L_{x^{-1}} g, \phi_S \rangle_S \qquad x \in G.$$

Then ψ is a bounded, continuous function on G and satisfies, on account of (13),

$$\psi(x\xi) = \psi(x) \text{ for } x \in G,\ \xi \in H;$$

that is,

$$(15) \qquad \psi = \dot{\psi} \circ \pi_H,$$

where $\dot{\psi}$ is a bounded, continuous function on the quotient space G/H. Now consider any function

$$(16) \qquad f_o \in [J^1(G,H)]^* \cap S^1(G).$$

By Lemma 1 we have for f_o and any $g \in S^1(G)$, on using (14),

$$\langle f_o * g, \phi_S \rangle_S = \int f_o(x)\psi(x^{-1})dx = \int \overline{f_o^*}(x)\psi(x)dx.$$

Applying the extended formula of Mackey-Bruhat (Ch. 8, §§ 2.1-2.3), we obtain, in view of (15):

$$\langle f_o * g, \phi_S \rangle_S = \int_{G/H} T_{H,q}\overline{f_o^*}(\dot{x}).\dot{\psi}(\dot{x})\, d_q\dot{x} = 0,$$

since $T_{H,q}\overline{f_o^*} = \overline{T_{H,q} f_o^*} = 0$ for any f_o in $[J^1(G,H)]^*$ (cf. (16)). Thus we have shown: if f_o satisfies (16), then for any $\phi_S \in S^1(G)'$ vanishing on $D_H S^1(G)$ we have

$$(17) \qquad \langle f_o * g, \phi_S \rangle_S = 0 \quad \text{for all } g \in S^1(G).$$

Now, if $S^1(G)$ is symmetric or pseudosymmetric, then $S^1(G)$ has approximate right units (§ 8, Proposition 1, (ii) and (iii)). Thus we can choose g in (17) so that $f_o * g$ is arbitrarily close to f_o (in $S^1(G)$); hence we obtain

$$\langle f_o, \phi_S \rangle_S = 0.$$

This implies $f_o \in D_H S^1(G)$, as the functional ϕ_S vanishing on $D_H S^1(G)$ was arbitrary. Since f_o was any function satisfying (16), we have obtained the inclusion opposite to (10). Thus (4) is proved. The proof of (5) is essentially the same - only simpler to write down - and may be left to the interested reader. This completes the second proof of Proposition 2.

In the particular case $S^1(G) = L^1(G)$, the proof just given reduces to a proof of the relations § 1, (34), (33); also, this proof of Proposition 2, by the method of linear functionals, is independent of Proposition 1 which may now be considered as a corollary of Proposition 2.

The method of linear functionals appears here as merely another method of proof; however, we shall see that this method also yields results which have not yet been obtained by other methods (cf. § 16, Theorem 1).

We also observe that this section affords another instance where Remark 5 of § 10 is pertinent.

§ 12. Segal algebras and normal subgroups

Let us now see what the results of the preceding sections yield for the case of normal subgroups.

(i) If $S^1(G)$ is symmetric or pseudosymmetric, then, when H is a normal subgroup, $D_H S^1(G)$ is a closed two-sided ideal of $S^1(G)$ and

$$(1) \qquad D_H S^1(G) = J^1(G,H) \cap S^1(G).$$

This follows from § 11, Prop. 1, and § 1, Corollary of Prop. 1, on the one hand, and from § 11, Proposition 2, and § 1, (22) on the other hand.

(ii) If $S^1(G)$ is symmetric, then for normal H we have

$$(2) \qquad D'_H S^1(G) = D_H S^1(G).$$

Also, if $S^1(G)$ is symmetric and stable under involution (thus a fortiori if $S^1(G)$ is star-symmetric), then for normal H we have

$$(3) \qquad [D_H S^1(G)]^* = D_H S^1(G).$$

(2) follows from § 11, Proposition 1, and § 1, Proposition 1; then we can obtain (3) by combining (2) with § 11, Proposition 2.

(iii) Let $S^1(G)$ be a symmetric Segal algebra. If H is a closed normal subgroup of G having the property P_1, then $D_H S^1(G)$ has multiple approximate left units and multiple approximate right units, bounded in L^1-norm by the constant 2.

This follows from § 10, Theorem 1, and (2) above.

(iv) Let $S^1(G)$ be symmetric; let H be a closed normal subgroup of G. If $D_H S^1(G)$ has approximate left or right units bounded, respectively, in the left or right operator norm of $D_H S^1(G)$, then H has the property P_1.

This follows from § 10, Theorem 2, combined with (2) above; cf.

also § 10, Remark 1.

(v) Let $S^1(G)$ be symmetric and let H be a closed normal subgroup of G. Then $D_H S^1(G)$ has approximate left [right] units bounded in L^1-norm, or in the left [right] operator norm of $D_H S^1(G)$, if and only if H has the property P_1. Moreover, the same holds true for multiple approximate left [right] units, bounded in L^1-norm or in the corresponding operator norm.

This is simply a combination of (iii) and (iv).

REMARK 1. This result shows in particular that, if $S^1(G)$ is symmetric and H is normal, then for $D_H S^1(G)$ the existence of approximate left [right] units bounded in the left [right] operator norm of $D_H S^1(G)$ is equivalent to the existence of multiple approximate left [right] units bounded in the L^1-norm. - This equivalence does not involve the property P_1; on the other hand, it is not at all obvious how it could be established without recourse to the properties P_1 (or P_*) and (M), even in the simplest case $S^1(G) = L^1(G)$, H = G. Whether the equivalence still persists if the subgroup H is not normal, is another open question.

For star-symmetric Segal algebras we can sharpen (iii) considerably.

(vi) Let $S^1(G)$ be a star-symmetric Segal algebra and H a closed, normal subgroup of G. If H has the property P_1, then the closed two-sided ideal $D_H S^1(G)$ has multiple approximate two-sided units which are self-adjoint and bounded in L^1-norm by the constant 2; more precisely, the following holds. Let

$$(4) \qquad (u_\alpha)_{\alpha \in A}$$

be multiple approximate left units of $S^1(G)$ such that

$$(5a) \quad u_\alpha^* = u_\alpha \qquad (5b) \quad \|u_\alpha\|_1 = 1$$

(cf. § 8, Proposition 1, part (i)). Put

(6) $$v_{\alpha,L} = u_\alpha * u_\alpha - (Lu_\alpha)^* * Lu_\alpha \quad \alpha \in A,\ L \in \mathcal{L}_H,$$

where $\mathcal{L}_H$ denotes the set of all operators L of the form

(7) $$L = \Sigma_n c_n L_{\xi_n},\ c_n > 0,\ \Sigma_n c_n = 1,\ \xi_n \in H.$$

Then $(v_{\alpha,L})$ constitutes a family of multiple approximate two-sided units in $D_H S^1(G)$ which are self-adjoint and bounded in L^1-norm by the constant 2.

The proof of (vi) is a refinement of that of § 10, Theorem 1, taking into account the stronger properties available.

That $D_H S^1(G)$ is here a two-sided ideal is already included in (i) above; but even (3) holds, since by assumption $S^1(G)$ is star-symmetric (and H is normal).

Now let $f_j \in D_H S^1(G)$, $1 \leq j \leq N$, and $\varepsilon > 0$ be given. Since H is a P_1-group, there is an operator $L \in \mathcal{L}_H$ such that

(8) $$\| Lf_j \|_S < \varepsilon \qquad j = 1,\dots,N.$$

(see § 10, (1), (2), and also § 10, Remark 2). Next, by § 8, Proposition 1, part (i), there is a $u \in S^1(G)$ such that

(9) $$\| u * f_j - f_j \|_S < \varepsilon \qquad j = 1,\dots,N$$

and

(10a) $\| u \|_1 = 1$ (10b) $u^* = u$

(we have dropped here the subscript α used in (4) and (5)). Now let us combine (8) and (9): since $L(g*h) = (Lg)*h$ and L is a contraction on $S^1(G)$, we obtain

(11) $$\| (Lu) * f_j \|_S < 2\varepsilon,$$

as in § 10, (7). From (11) we obtain by left convolution with $(Lu)^*$

(12) $$\| (Lu)^* * (Lu) * f_j \|_S < 2\varepsilon,$$

since $\|(Lu)^*\|_1 \leqslant \|u\|_1 = 1$ (cf. (10a)). Next we note that (9) implies, again by (10a),

$$(13) \qquad \|u * u * f_j - f_j\|_S < 2\varepsilon.$$

Consider the difference of the expressions in (13) and (12): we can write

$$(14) \qquad u * u * f_j - f_j - (Lu)^* * (Lu) * f_j = v * f_j - f_j,$$

where

$$(15) \qquad v = u * u - (Lu)^* * (Lu).$$

Thus (14), (13), (12) show that

$$(16) \qquad \|v * f_j - f_j\|_S < 4\varepsilon \qquad j = 1,\ldots,N.$$

Clearly v satisfies (cf. (15) and (10b))

$$(17) \qquad v^* = v.$$

Moreover, for the L^1-norm of v we have by (10a)

$$(18) \qquad \|v\|_1 \leqslant \|u\|_1 . \|u\|_1 + \|u\|_1 . \|u\|_1 = 2.$$

Let us show that v lies in $D_H S^1(G)$: indeed, by (10b) we can write (15) in the form

$$(19) \qquad v = u * (u - Lu) + (u - Lu)^* * Lu.$$

Recalling the meaning of L (cf. (7)), we can write

$$u - Lu = \Sigma_n c_n . (u - L_{\xi_n} u),$$

which shows that u - Lu lies in $D_H S^1(G)$, since $\xi_n \in H$; thus by (3) $(u - Lu)^*$ is also in $D_H S^1(G)$. As $D_H S^1(G)$ is a two-sided ideal of $S^1(G)$, (19) now shows:

(20) $$v \in D_H S^1(G).$$

Finally, it is easy to see that the functions of the type (15) are, in fact, approximate two-sided units for $D_H S^1(G)$. Indeed, given f_j, $1 \leqslant j \leqslant N$, in $D_H S^1(G)$ and $\varepsilon > 0$, we may consider the 2N functions f_j, f_j^* (the latter being also in $D_H S^1(G)$!): by what we have shown already, there is a v of the form (15), and thus satisfying (17), (18), (20), such that

$$\|v * f_j - f_j\|_S < \varepsilon \quad \text{and} \quad \|v * f_j^* - f_j^*\|_S < \varepsilon \qquad j = 1,\dots,N.$$

Then we get by involution and in view of (17)

$$\|v * f_j - f_j\|_S < \varepsilon \quad \text{and} \quad \|f_j * v - f_j\|_S < \varepsilon \qquad j = 1,\dots,N.$$

Thus (vi) is proved.

REMARK 2. In the proof of (vi) a crucial part was played by the fact that $u^* = u$ (cf. (10b)). One might at first attempt to do without this, replacing (15) by $v = u^* * u - (Lu)^* * (Lu)$. But (9) does just yield (13), which involves $u * u$ and not $u^* * u$! On the other hand, if we put $v = u * u - (Lu) * (Lu)$, we lose (17), even if $u^* = u$. Note also the part played by the fact that here $D_H S^1(G)$ is stable under involution; the isometry of the involution is used only at the very end of the proof.

(vii) As a particular case of (vi) we have for $S^1(G) = L^1(G)$: if H is a closed, normal subgroup of G with the property P_1, then the functions

$$u * u - (Lu)^* * (Lu),$$

where u ranges over all functions in $\mathcal{K}_+(G)$ such that $\int u = 1$ and $u^* = u$, and L over $\mathcal{L}_H$ (cf. (7)), form a family of approximate two-sided units in $D_H L^1(G)$ which are bounded in L^1-norm by the constant 2 and are self-adjoint.

(viii) Let G be a l.c. abelian group, H a closed subgroup; then, if $S^1(G)$ is star-symmetric, the following holds: the ideal $D_H S^1(G)$ has multiple approximate units which are bounded in L^1-norm by the constant 2 and have a positive Fourier transform with compact support.

This follows at once from (vi): for, it is readily verified that, in the case of abelian G, the functions $v_{\alpha,L}$ defined in (6) have a positive Fourier transform; they also have L^1-norm $\leqslant 2$ (cf. (18)). Now consider the functions $v_\gamma * v_{\alpha,L}$, where v_γ ranges over all (continuous) functions in $L^1(G)$ of L^1-norm 1 that have a positive Fourier transform with compact support: the functions $v_\gamma * v_{\alpha,L}$ are arbitrarily close to $v_{\alpha,L}$ by § 8, Proposition 2 (it is even enough here to interpret 'arbitrarily close' only in the sense of $L^1(G)$). Since v_γ belongs to $S^1(G)$ (Ch. 6, § 2.2, (iii)), the functions $v_\gamma * v_{\alpha,L}$ are in $D_H S^1(G)$; they clearly constitute a family of multiple approximate units for $D_H S^1(G)$, of L^1-norm $\leqslant 2$, having a positive Fourier transform with compact support.

A generalization of (viii) will be established in § 17, Theorem 2 (cf. also § 17, Remark 3).

The results given in this section certainly need clarification by further research; this applies in particular to Remark 1 above.

Note. For general l.c. groups G, the functions (6) - see also (5a) and (7) - define positive operators on $L^2(G)$ by (left) convolution, as is readily seen; for abelian G this is equivalent to the positivity of the Fourier transforms.

§ 13. Segal algebras on quotient groups

Let H be closed, normal subgroup of G. It can be shown in a very simple way that the image of $L^1(G)$ under the mapping T_H (cf. § 1, (16)) is $L^1(G/H)$ and that

(1) $$L^1(G/H) \cong L^1(G)/J^1(G,H),$$

the isomorphism being not only algebraic, but also isometric (Ch. 3, §§ 4.4 and 5.3). Further, if H has the property P_1, then it can also be proved that the image of a closed left [right] ideal of $L^1(G)$ under the mapping T_H is a closed left [right] ideal of $L^1(G)$ (Ch. 8, § 4.6 (ii)). We establish here analogous results for Segal algebras.

THEOREM 1. Let $S^1(G)$ be a Segal algebra and H a closed, normal subgroup of G. Let $S^1(G/H)$ be the image of $S^1(G)$ under the mapping T_H: $L^1(G) \to L^1(G/H)$. Then the following holds.

(i) $S^1(G/H)$ is a Segal algebra under the quotient norm:

(2) $$S^1(G/H) \cong S^1(G)/J^1_S(G,H),$$

where $J^1_S(G,H)$ is the kernel of the restriction of T_H to $S^1(G)$,

(3) $$J^1_S(G,H) = \{f \mid f \in S^1(G), T_Hf = 0\} = J^1(G,H) \cap S^1(G).$$

(ii) If $S^1(G)$ is symmetric, pseudosymmetric or star-symmetric, then $S^1(G/H)$ is also symmetric, pseudosymmetric or star-symmetric, respectively.

(iii) If $S^1(G)$ is symmetric and H has the property P_1, then the image under T_H of a closed left or right ideal of $S^1(G)$ is a closed left or right ideal of $S^1(G/H)$, respectively; for left ideals this also holds if $S^1(G)$ is pseudosymmetric.

REMARK. The following distinction between (1) and (2) should be noted. In (1) we have two Banach algebras, defined independently of one another. But in (2) the norm for $S^1(G/H)$ is by definition that of $S^1(G)/J^1_S(G,H)$; the actual determination of this norm in $S^1(G/H)$, in terms of $T_H f \in S^1(G/H)$ alone, independently of the norm in $S^1(G)$, is another question.

Proof of Theorem 1.

(i) By the linearity of T_H, $S^1(G/H)$ is a linear subspace of $L^1(G)$. Let us verify that $S^1(G/H)$ satisfies the condition $S_0,\dots,S_3$ of § 4. Clearly $S^1(G/H)$ is dense in $L^1(G/H)$, by § 1, (18), since $S^1(G)$ is dense in $L^1(G)$. Next we have by definition

$$(4) \qquad \|T_H f\|_{S^1(G/H)} = \inf_{f_0} \|f + f_0\|_S \quad (f_0 \in J^1_S(G,H)).$$

Replacing on the right the Segal norm by the $L^1(G)$-norm, and $J^1_S(G,H)$ by $J^1(G,H)$, we obtain

$$\|T_H f\|_{S^1(G/H)} \geqslant \inf_{f_0} \|f + f_0\|_1 \quad (f_0 \in J^1(G,H))$$
$$\geqslant \|T_H f\|_{L^1(G/H)}$$

(cf. § 1, (18), (21); in fact the two terms on the right are equal, but this is not essential here). That $S^1(G/H)$, with the quotient norm (4), is a Banach space (and even a Banach algebra, since $S^1(G)$ is a Banach algebra), is well-known. Also, $S^1(G)$ is left invariant, since for $\dot{a} \in G/H$, say $\dot{a} = \pi_H(a)$, $a \in G$, we have

$$(5) \qquad T_H L_a f = L_{\pi_H(a)} T_H f \qquad f \in L^1(G).$$

The continuity of $\dot{y} \to L_{\dot{y}}\dot{f}$ ($\dot{f} \in S^1(G/H)$) follows readily from (5) and the continuity of $y \to L_y f$ ($f \in S^1(G)$), for we have

$$(6) \qquad \|T_H L_a f - T_H L_b f\|_{S^1(G/H)} \leqslant \|L_a f - L_b f\|_S \quad a,b \in G.$$

Finally, the left invariance of the norm (4) results from the left invariance of the norm in $S^1(G)$ and the fact that $J^1_S(G,H)$ is left invariant (since $S^1(G)$ and $J^1(G,H)$ are). Thus $S^1(G/H)$, with the norm (4), satisfies all conditions for a Segal algebra.

(ii) Now suppose that $S^1(G)$ satisfies the condition S_2' of § 4. Then $S^1(G/H)$ is right invariant, in virtue of the formula

$$(7) \qquad T_H R_a f = R_{\pi_H(a)} T_H f \qquad a \in G,\ f \in L^1(G),$$

analogous to (5); here, of course, $R_{\pi_H(a)}$ is defined by

$$(8) \qquad R_{\dot{a}} \dot{f}(\dot{x}) = \dot{f}(\dot{x}\dot{a}^{-1})\Delta_{G/H}(\dot{x}^{-1}) \quad \dot{a} \in G/H,\ \dot{f} \in L^1(G/H).$$

(8) can be proved by reduction to (5): we first use § 1, (9) and § 1, (20); then we apply (5), use § 1, (20) again, and finally apply the analogue of § 1, (9) on G/H. The continuity of $\dot{y} \to R_{\dot{y}}\dot{f}$, $\dot{y} \in G/H$ ($f \in S^1(G/H)$) follows from (7) and the continuity of $y \to R_y f$ $y \in G$ ($f \in S^1(G)$), via the 'symmetric' analogue of (6).

The condition S_3' of § 4, i.e. the invariance of the norm in $S^1(G)$ under R_y ($y \in G$), entails at once that the quotient norm (4) in $S^1(G/H)$ is invariant under $R_{\dot{y}}$ ($\dot{y} \in G/H$), by (7) (cf. also (8)) and the right invariance of $J^1_S(G,H)$.

If $S^1(G)$ satisfies the condition S^0 of § 4, then clearly so does $S^1(G/H)$.

Thus we have shown that, if $S^1(G)$ is symmetric or pseudosymmetric, then so is $S^1(G/H)$. Now let $S^1(G)$ be star-symmetric: then the formula § 1, (20) shows, first, that $S^1(G/H)$ is stable under involution in $L^1(G/H)$ and, secondly, that the norm (4) is invariant under this involution (note that $J^1_S(G,H)$ is stable under the involution of $L^1(G)$, since $S^1(G)$ and $J^1(G,H)$ are); thus $S^1(G/H)$ is star-symmetric.

(iii) Here the basic tool for the proof is this: if $S^1(G)$ is <u>symmetric or pseudosymmetric</u>, then the equality

(9) $$D_H S^1(G) = J_S^1(G,H)$$

holds (cf. § 12, (i)). Thus we may apply Glicksberg's theorem (Ch. 8, § 6.1) to the operators $(L_\xi)_{\xi\in H}$, since by hypothesis H satisfies P_1 and hence (M); this yields the relation

(10) $$\inf_{L\in\mathcal{L}_H} \|Lf\|_S = \|T_H f\|_{S^1(G/H)} \qquad f \in S^1(G),$$

in view of (2) and (9). Now any closed left ideal I of $S^1(G)$ is left invariant (§ 9, Proposition 1). The proof that $T_H(I)$ is closed in $S^1(G/H)$ is then exactly the same, by means of (10), as that given for the case of $L^1(G)$ in Ch. 8, § 4.6 (i), only with a change in notation. Also $T_H(I)$ is clearly a left ideal of $S^1(G/H)$. For closed right ideals we note that, if $S^1(G)$ is symmetric, then we also have, on combining (9) with § 12 (ii),

(11) $$D'_H S^1(G) = J_S^1(G,H).$$

We now apply Glicksberg's theorem to $(R_{\xi^{-1}})_{\xi\in H}$ (we take $R_{\xi^{-1}}$ here, since $R_{ab} = R_b R_a$!); this yields

(12) $$\inf_{R\in\mathcal{R}_H} \|Rf\|_S = \|T_H f\|_{S^1(G/H)} \qquad f \in S^1(G).$$

The proof that $T_H(I)$ is closed in $S^1(G/H)$ if I is a closed right ideal of $S^1(G)$ is then the same as before, since I is right invariant (§ 9, Proposition 1).

It should be observed that in the case of $L^1(G)$ we can transform a left ideal into a right one, and vice-versa, by involution (cf. Ch.8, § 4.6 (ii)); but we cannot do this for symmetric Segal algebras, as we do not know whether symmetry implies star-symmetry.

Thus the proof of Theorem 1 is complete.

It may be mentioned that, if $S^1(G)$ is symmetric and H is a closed, normal subgroup of G with the property P_1, then for every closed left or right ideal I of $L^1(G)$ the relation

(13) $$T_H(I \cap S^1(G)) = T_H(I) \cap T_H(S^1(G))$$

holds; for left ideals this is also true if $S^1(G)$ is pseudosymmetric. (13) is a corollary of Theorem 1 and § 9, Theorem 1.

REMARK 1. Theorem 1 can be established without recourse to Glicksberg's theorem.

Proof. We want to give a new proof of part (iii) of Theorem 1. For this, we recall the results § 2, (9) and (10), noting also that $[J^1(G,H)]^*$ and $J^1(G,H)$ coincide in our case (§ 1, (22)). From these results relating to $L^1(G)$ we can obtain their analogues for a symmetric Segal algebra $S^1(G)$ (and a normal P_1-subgroup H):

(14) If $f \in J^1_S(G,H)$, then $\inf_{L\in\mathcal{L}_H} \|Lf\|_S = 0$.

(15) If $f \in J^1_S(G,H)$, then $\inf_{R\in\mathcal{R}_H} \|Rf\|_S = 0$.

Here (14) also holds if $S^1(G)$ is pseudosymmetric.

The method for deducing (14) and (15) from § 2, (9) and (10), respectively, is just the same as that used for proving § 10, Remark 3.

We note that by this method of proof we obtain (14) from § 2, (9) without any appeal to (9) above; likewise we obtain in this way (15) from § 2, (10) without making use of (11). This is of interest in connection with the fact that (9) follows from (14): for, we can write for $f \in S^1(G)$

$$f - (f - Lf) = Lf$$

and here $f - Lf$ lies in $D_H S^1(G)$, hence (14) implies for $f \in J^1_S(G,H)$ that f lies in $D_H S^1(G)$; thus (9) clearly follows. Likewise (11) follows from (15).

Now we shall show that (14) implies (10), and (15) implies (12); thereafter the proof of part (iii) of Theorem 1 proceeds as indicated before.

The implications just stated are best established in a general setting as follows.

LEMMA. Let $(A_\xi)_{\xi\in H}$ be <u>any</u> family of linear contraction operators on a Banach space B (with norm $\|.\|$). Let $D_H B$ be the closed linear subspace generated by the set

$$\{f - A_\xi f \mid f \in B, \xi \in H\}.$$

Let $\mathcal{A}_H$ consist of all operators

$$A = \Sigma_n c_n A_{\xi_n} \quad \text{with } c_n > 0,\ \Sigma_n c_n = 1,\ \xi_n \in H$$

(i.e. $\mathcal{A}_H$ is the convex hull of the given family). Suppose that

$$(16) \qquad \inf_{A\in\mathcal{A}_H} \|Af_o\| = 0 \ \underline{\text{for all}}\ f_o \in D_H B$$

(note that if this holds for $f \in B$, then necessarily $f \in D_H B$; cf. (18) below). Then we have

$$(17) \qquad \inf_{A\in\mathcal{A}_H} \|Af\| = \|T_H f\|_{B/D_H B} \text{ for all } f \in B,$$

where $T_H f$ is the canonical mapping of B onto the quotient space $B/D_H B$.

<u>Proof</u>. Consider any $A \in \mathcal{A}_H$. For any $f \in B$ we can write

$$Af = f - (f - Af),$$

and here $f - Af = \Sigma_n c_n (f - A_{\xi_n} f)$ lies in $D_H B$, hence we have

$$(18) \qquad \|Af\| \geqslant \|T_H f\|_{B/D_H B} \text{ for all } A \in \mathcal{A}_H.$$

On the other hand, given $\varepsilon > 0$, there is an $f_o \in D_H B$ such that

$$\|f - f_o\| < \|T_H f\|_{B/D_H B} + \varepsilon.$$

By (16) there is an $A' \in \mathcal{A}_H$ such that $\|A'f_o\| < \varepsilon$. Now we can write by the linearity of A'

$$A'f = A'(f - f_o) + A'f_o,$$

whence, since A' is a contraction,

(19) $\|A'f\| \leq \|f - f_o\| + \|A'f_o\| < \|T_H f\|_{B/D_H B} + 2\varepsilon.$

(19) and (18) yield (17), and the Lemma is proved.

This concludes the proof of Remark 1. The essential feature of the proof is that relations (10) and (12) can be established, for the appropriate class of Segal algebras and normal P_1- subgroups, by a 'reduction to $L^1(G)$'; cf. in this respect Remark 5 at the end of § 10.

The role which the property P_1 plays in part (iii) of Theorem 1 needs to be clarified; this was already observed, in the case of L^1-algebras, in Ch. 8, § 4.6.

§ 14. Closed ideals with approximate units in Segal algebras

In § 13 we studied 'Segal quotient algebras' $S^1(G/H)$ of a Segal algebra $S^1(G)$. Our previous work will permit us to show that, if we know closed ideals with approximate units in $S^1(G/H)$, we can construct closed ideals of $S^1(G)$ that have approximate units, under appropriate conditions.

First we formulate two general theorems.

THEOREM 1. Let $S^1(G)$ be a Segal algebra and H a closed normal subgroup of G with the property P_1. Consider the Segal quotient algebra $S^1(G/H)$, the image of $S^1(G)$ under the mapping T_H (§ 13, Theorem 1). Let A be any subalgebra of $S^1(G/H)$ and define the algebra $B \subset S^1(G)$ by

$$B = \{f \mid f \in S^1(G), T_H f \in A\}.$$

If $S^1(G)$ is symmetric or pseudosymmetric, then the following holds: whenever A has [multiple] approximate left units, then B has [multiple] approximate left units; if $S^1(G)$ is symmetric then this implication also holds for [multiple] approximate right units.

Proof. Here several previous results are brought together. A and B are normed algebras with norms induced by $S^1(G/H)$ and $S^1(G)$, respectively; moreover, we may write

$$A \cong B/J^1_S(G,H),$$

by the definition of the norm in $S^1(G/H)$ (cf. § 13, Theorem 1). Now $J^1_S(G,H)$ coincides with $D_H S^1(G)$, for any normal subgroup H, if $S^1(G)$ is symmetric or pseudosymmetric (§ 12, (i)). $D_H S^1(G)$ has multiple approximate left units, for any $S^1(G)$ and any closed P_1-subgroup H (§ 10, Theorem 1); $D_H S^1(G)$ also has multiple approximate right units whenever $S^1(G)$ is symmetric and H is a closed normal P_1-subgroup

(§ 12, (iii)). These approximate left or right units are bounded in L^1-norm by the constant 2 (cf. again § 10, Theorem 1, and § 12 (iii)), thus they are a fortiori bounded in the corresponding operator norm of B. Hence we can apply § 7, Lemma 2, (cf. also Remarks 1 and 2 at the end of § 7) to B and $J_S^1(G,H)$; this yields Theorem 1.

THEOREM 2. Let $H \subset G$ be a closed normal subgroup with the property P_1. Let A be a subalgebra of $L^1(G/H)$ and define the subalgebra B of $L^1(G)$ by $B = T_H^{-1}(A)$. If A has [multiple] left or right units bounded in the $L^1(G/H)$-norm, then B has [multiple] approximate left or right units, respectively, bounded in the $L^1(G)$-norm.

<u>Proof</u>. This is entirely analogous to the proof of Theorem 1 above. We have here, on providing the algebra A with the $L^1(G/H)$-norm and the algebra B with the $L^1(G)$-norm,

$$A \cong B/J^1(G,H)$$

(cf. Ch. 3, §§ 4.4 and 5.3). $J^1(G,H)$ - which coincides with $D_H L^1(G)$ if H is normal (§ 1, (24)) - has multiple approximate left units <u>and</u> multiple approximate right units, bounded in the $L^1(G)$-norm, if H is a normal P_1-subgroup (cf. § 12, (iii)). We can now apply § 7, Lemma 3 (cf. also the Remarks 1 and 2 at the end of § 7) to B and $J^1(G,H)$ which yields Theorem 2.

In order to apply these theorems, we now give some definitions. We consider any Banach algebra B (in the applications we will have B = = $S^1(G)$ or $B = L^1(G)$) and introduce various families of closed, <u>two-sided</u> ideals of B, as follows:

(i) $\mathcal{I}_a(B)$ [$\mathcal{I}_{ma}(B)$] denotes the family of all closed, two-sided ideals I of B with [multiple] approximate left units: the statement $I \in \mathcal{I}_a(B)$ [$I \in \mathcal{I}_{ma}(B)$] means that I, considered as a Banach algebra of its own, satisfies the condition (a) [(ma)] of § 6.

(ii) $\mathcal{I}_b(B)$ [$\mathcal{I}_{mb}(B)$] denotes the family of all closed, two-sided ideals I of B with bounded [multiple] approximate left units, i.e. all ideals I which, as Banach algebras, satisfy condition (b) [(mb)] of § 6.

(iii) Replacing in (i) and (ii) 'left' by 'right', we define the families

$$\mathcal{I}_{a'}(B),\ \mathcal{I}_{ma'}(B),\ \mathcal{I}_{b'}(B),\ \mathcal{I}_{mb'}(B)$$

of all closed, two-sided ideals of B satisfying, respectively, the conditions (a′), (ma′), (b′), (mb′) of § 6.

If we omit the word 'closed' in the definition of the families $\mathcal{I}_x(B)$ above, then these definitions retain a meaning, and Corollaries 1 and 2 below are still true; but in practice we always deal with closed ideals.

We note two <u>properties of the families</u> $\mathcal{I}_x(B)$:

1. Each of the families defined in (i), (ii), (iii) above contains with two ideals also their <u>intersection</u>.

Consider e.g. $I_1, I_2 \in \mathcal{I}_a(B)$. Let f be in $I_1 \cap I_2$, which is a closed, two-sided ideal. Given $\varepsilon > 0$, there is a $u_1 \in I_1$ such that

$$\|u_1.f - f\| < \varepsilon/2.$$

Since $u_1.f$ lies in I_2, there is a $u_2 \in I_2$ such that

$$\|u_2.(u_1.f) - u_1.f\| < \varepsilon/2.$$

Thus for $u = u_2.u_1$ we have:

$$u \in I_1 \cap I_2,\ \|u.f - f\| < \varepsilon.$$

Note that the proof depends on the fact that I_1 and I_2 are two-sided ideals.

2. Each of the families defined in (i), (ii), (iii) is invariant under an <u>automorphism</u> of B (by an automorphism of a Banach algebra B

is meant a bijective mapping of B onto itself which preserves the algebraic operations _and_ the norm).

We can now apply Theorems 1 and 2 to the case $B = S^1(G)$ or $L^1(G)$; in Corollaries 1 and 2 below the notations introduced above will be used, for conciseness.

COROLLARY 1. Let $S^1(G)$ be a symmetric or pseudosymmetric Segal algebra. Let _finitely many_ closed, normal subgroups H_n of G be given which have the property P_1, and consider the Segal algebras $S^1(G/H_n)$, the images of $S^1(G)$ under T_{H_n} (cf. Theorem 1). _If for each_ n

(1) $$I_n \in \mathcal{J}_a(S^1(G/H_n)) \quad [\mathcal{J}_{ma}(S^1(G/H_n))],$$

then

$$\bigcap_n T^{-1}_{H_n,S}(I_n),$$

where $T_{H_n,S}$ denotes the restriction of T_{H_n} to $S^1(G)$, _lies in_

(2) $$\mathcal{J}_a(S^1(G)) \quad [\mathcal{J}_{ma}(S^1(G))].$$

More generally, if for each n an automorphism α_n of $S^1(G)$ is given, then _this also holds for_

$$\bigcap_n \alpha_n(T^{-1}_{H_n,S}(I_n)).$$

If $S^1(G)$ is symmetric, the corresponding 'right-hand' version is also true, i.e. we may replace in (1) and (2) $\mathcal{J}_a$ by $\mathcal{J}_{a'}$, $\mathcal{J}_{ma}$ by $\mathcal{J}_{ma'}$.

Proof. This follows at once from Theorem 1 and the two properties of the families $\mathcal{J}_x$ indicated above.

COROLLARY 2. Let _finitely many_ closed, normal subgroups H_n of a l.c. group G be given which have the property P_1. _If for each_ n

(3) $$I_n \in \mathcal{J}_b(L^1(G/H_n)) \quad [\mathcal{J}_{mb}(L^1(G/H_n))],$$

<u>then</u>

$$\bigcap_n T_{H_n}^{-1}(I_n)$$

<u>lies in</u>

(4) $$\mathcal{I}_b(L^1(G)) \quad [\mathcal{I}_{mb}(L^1(G))].$$

More generally, <u>the same holds for</u>

(5) $$\bigcap_n \chi_n . T_{H_n}^{-1}(I_n),$$

where each χ_n is a character of G. The assertion remains true if we replace in (3) and (4) $\mathcal{I}_b$ by $\mathcal{I}_{b'}$, $\mathcal{I}_{mb}$ by $\mathcal{I}_{mb'}$ ('right-hand' version).

<u>Proof</u>. This follows from Theorem 2 and the properties of the families $\mathcal{I}_x$. Concerning (5) we note that $f \to \chi.f$ is an automorphism of $L^1(G)$ for any character of G (the definition of character includes continuity and applies to arbitrary topological groups; $\chi.I$ denotes the set $\{\chi.f \mid f \in I\}$).

Let us now consider some applications of these corollaries.

(I) First, let G be <u>abelian</u>.

(I_1) $\mathcal{I}_a(S^1(G))$ <u>contains</u>, for instance, <u>all closed ideals with countable cospectrum</u>, for any Segal algebra $S^1(G)$. We shall discuss this, and related matters, later (§ 16).

(I_2) Suppose there is a (non-trivial) idempotent measure $\mu \in M^1(G)$, i.e. $\mu * \mu = \mu$ ($\mu \neq 0, \delta_e$, cf. Ch. 3, § 5.5). Then

$$\mu * S^1(G) = \{\mu * f \mid f \in S^1(G)\}$$

belongs to $\mathcal{I}_{ma}(S^1(G))$, for any Segal algebra $S^1(G)$.

First, $\mu * S^1(G)$ is contained in $S^1(G)$ (cf. § 4, Proposition 2); moreover, $\mu * S^1(G)$ is a <u>closed</u> ideal of $S^1(G)$, since μ is idempotent: indeed, if

$$\mu * f_n \to g \quad (n \to \infty)$$

in $S^1(G)$, then

$$\mu * f_n = \mu * \mu * f_n \to \mu * g,$$

thus $g = \mu * g$ lies in $\mu * S^1(G)$. Now let any finite set F in $\mu * S^1(G)$ and $\varepsilon > 0$ be given. There is a $u \in S^1(G)$ such that

(6) $$\|u * f - f\|_S < \varepsilon/\|\mu\| \qquad \text{for each } f \in F$$

(cf. § 8, Proposition 1, (i)). Thus (cf. again § 4, Proposition 2)

$$\|\mu * u * f - \mu * f\|_S < \varepsilon \qquad f \in F.$$

Since $\mu = \mu * \mu$ and $f = \mu * g$, with $g \in S^1(G)$, we have $\mu * f = f$, whence

(7) $$\|(\mu * u) * f - f\|_S < \varepsilon \qquad f \in F,$$

and $\mu * u$ lies in $\mu * S^1(G)$.

(I_3) Suppose there is a non-trivial idempotent measure $\mu \in M^1(G)$. Then $\mu * L^1(G)$ belongs to $\mathcal{J}_{mb}(L^1(G))$.

Indeed, in the proof above (cf. (6), (7)) we may take u so that $\|u\|_1 = 1$; then $\mu * u$ in (7) satisfies $\|\mu * u\|_1 \leqslant \|\mu\|$, i.e. $\mu * u$ is bounded in $L^1(G)$.

(II) Secondly, let G be soluble and let (H_n) be a finite family of closed normal subgroups with abelian quotient groups. Given $S^1(G)$, put $S^1(G/H_n) = T_{H_n} S^1(G)$; we denote by $T_{H_n,S}$ the restriction of T_{H_n} to $S^1(G)$.

(II_1) If I_n lies in $\mathcal{J}_a(S^1(G/H_n))$, for each n, then

(8) $$\bigcap_n T^{-1}_{H_n,S}(I_n)$$

lies in $\mathcal{J}_a(S^1(G))$ and in $\mathcal{J}_{a'}(S^1(G))$, whenever $S^1(G)$ is symmetric; if $S^1(G)$ is pseudosymmetric, (8) still belongs to $\mathcal{J}_a(S^1(G))$.

This follows from Corollary 1 (note that $H_n \subset G$ is soluble, hence a P_1-group).

(II_2) Suppose there is a non-trivial idempotent measure $\mu_n \in M^1(G/H_n)$, for each n. Then

$$\bigcap_n T^{-1}_{H_n,S}(\mu_n * S^1(G/H_n)) \tag{9}$$

<u>lies in</u> $\mathcal{T}_{ma}(S^1(G))$ <u>and in</u> $\mathcal{T}_{ma'}(S^1(G))$, <u>whenever</u> $S^1(G)$ <u>is symmetric;</u> if $S^1(G)$ is pseudosymmetric, (9) still belongs to $\mathcal{T}_{ma}(S^1(G))$.

This follows from (I_2) and Corollary 1.

(II_3) Suppose there is a non-trivial idempotent measure $\mu_n \in M^1(G/H_n)$, for each n. Then

$$\bigcap_n \chi_n \cdot T^{-1}_{H_n}(\mu_n * L^1(G/H_n)), \tag{10}$$

where each χ_n is a character of G, <u>lies in</u> $\mathcal{T}_{mb}(L^1(G))$ <u>and in</u> $\mathcal{T}_{mb'}(L^1(G))$.

This follows from (I_3) and Corollary 2.

EXAMPLES

1. $G = \mathbb{R}$, $H_n = \lambda_n\mathbb{Z}$ ($\lambda_n > 0$); the group $\mathbb{R}/\lambda_n\mathbb{Z} \cong \mathbb{R}/\mathbb{Z}$ carries non-trivial idempotent measures, as is well-known.

2. G is the multiplicative group of all matrices

$$\begin{pmatrix} e^t & x \\ 0 & 1 \end{pmatrix} \qquad t \in \mathbb{R},\ x \in \mathbb{R}. \tag{11}$$

H_n is the closed, normal subgroup of all matrices (11) with t restricted to $\lambda_n\mathbb{Z}$, for $\lambda_n > 0$. Then $G/H_n \cong \mathbb{R}/\lambda_n\mathbb{Z}$.

3. More generally, let G_1 be a l.c. abelian group and H'_n closed subgroups such that G_1/H'_n is compact. Consider a semi-direct product G of G_1 with another l.c. abelian group G_2: thus G contains G_2 as closed normal subgroup and $G/G_2 = G_1$. Let H_n be the inverse image of H'_n in G; then $G/H_n \cong G_1/H'_n$ (cf. WEIL [1, p. 12]) is a compact abelian

group, hence carries non-trivial idempotent measures.

In this section we have constructed some closed ideals in $S^1(G)$, or in $L^1(G)$, with various kinds of approximate units; we shall later utilize these constructions again.

The various families of ideals introduced in this section occur, as it were, in nature; this is the main reason for the general definition of approximate units introduced in § 6. In this context we may also refer to the results in §§ 16 and 17.

As to the groups considered in this section, we have dealt here with abelian and with soluble groups. It is natural to study the analogous situation arising in the case of compact groups and motion groups; this case will be considered next (§ 15).

§ 15. Symmetric Segal algebras on compact groups

THEOREM 1. Let $S^1(G)$ be a symmetric Segal algebra on a compact group G. Then every closed, two-sided ideal of $S^1(G)$ contains multiple approximate two-sided units u of the form

$$u = \Sigma_n c_n \chi_{D_n}, \quad c_n > 0.$$

That is, u is a (finite) linear combination of characters χ_{D_n} of finite-dimensional irreducible representations D_n of G, with positive coefficients c_n; thus these approximate units u are positive-definite, in particular self-adjoint, and lie in the centre of $L^1(G)$.

Proof. Let I be a closed two-sided ideal of $S^1(G)$; let any finite set $F \subset I$ and $\varepsilon > 0$ be given. Let U be a nd. of $e \in G$ such that

$$\|L_y f - f\|_S < \varepsilon/2 \quad \text{for all } f \in F.$$

We recall that, if $u_1 \in \mathcal{K}(G)$ - not necessarily in $S^1(G)$ - is such that $u_1 \geqslant 0$, $\int u_1 = 1$ and Supp $u_1 \subset U$, then

(1) $$\|u_1 * f - f\|_S < \varepsilon/2 \qquad f \in F.$$

(cf. the first part of the proof of § 8, Proposition 1, (i)). Now let V be a nd. of e so small that $V.V^{-1} \subset U$. There is a $v \in \mathcal{K}(G)$ such that $v \geqslant 0$, $\int v = 1$, Supp $v \subset V$ and which lies in the centre of $L^1(G)$: this is a fundamental fact for which we refer to WEIL [1, p. 86]. We may take in (1)

$$u_1 = v * v^*,$$

since Supp $v * v^* \subset V.V^{-1}$; thus

(2) $$\|v * v^* * f - f\|_S < \varepsilon/2 \qquad f \in F.$$

Let v have the 'Fourier series' (WEIL [1, p. 85])

(3) $$\Sigma_D \hat{v}_D . \chi_D,$$

where D ranges over all equivalence classes of finite-dimensional irreducible representations of G, χ_D is the <u>character</u> of the representation (of class) D, and the 'Fourier coefficient' $\hat{v}_D$ is given by

(4) $$\hat{v}_D = \int v(x)\overline{\chi_D(x)}dx.$$

We recall that χ_D is continuous, $\chi_D(e)$ = dim D (the dimension of D) and χ_D <u>lies in the centre of</u> $L^1(G)$; moreover

(5) $$\chi_D * \chi_D = (1/\dim D).\chi_D.$$

We have here

$$\chi_D^*(x) = \overline{\chi_D(x^{-1})},$$

since G is unimodular. The character χ_D is a <u>positive-definite function</u> in particular

$$\chi_D^* = \chi_D.$$

The function $v * v^*$ in (2) has the Fourier series

(6) $$\Sigma_D |\hat{v}_D|^2.(\dim D)^{-1}.\chi_D$$

(cf. WEIL [1], p. 86, where on line 12 from below in the last term the factor r should be omitted). The series (6) converges to $v * v^*$ in $L^2(G)$ (WEIL [1], p. 85); it even converges uniformly (loc.cit., p. 77), but this fact is not needed here. Thus there is a <u>finite</u> linear combination, say

(7) $$w = \Sigma_n c_n \chi_{D_n},$$

with coefficients

(8) $$c_n > 0,$$

such that, if

$$M = \max\|f\|_S \ (f \in F),$$

the approximation

$$\|v * v^* - w\|_2 < \varepsilon/(2M)$$

holds in $L^2(G)$. For the $L^1(G)$-norm we then have, by Schwarz's inequality, the same approximation:

$$(9) \qquad \|v * v^* - w\|_1 < \varepsilon/(2M).$$

Observing that in $S^1(G)$ the estimate

$$\|w * f - f\|_S \leqslant \|w - v * v^*\|_1 . \|f\|_S + \|v * v^* * f - f\|_S$$

holds, we obtain from (9) and (2)

$$(10) \qquad \|w * f - f\|_S < \varepsilon \qquad f \in F.$$

Now let us put

$$(11) \qquad u = \Sigma_{(n)} c_n \chi_{D_n},$$

<u>where the summation is extended over those and only those values of</u> n <u>in the finite sum</u> (7) <u>for which</u>

$$(12) \qquad \chi_D * f \neq 0 \quad \text{for at least } \underline{\text{one}} \ f \in F,$$

i.e. (11) is obtained from (7) by omitting all those terms $c_n . \chi_{D_n}$ for which $\chi_{D_n} * f = 0$ for <u>each</u> $f \in F$, Then, of course,

$$(13) \qquad u * f = w * f \qquad f \in F.$$

(but this 'of course' embodies the main idea of the proof, jointly with the use of a fundamental fact at the beginning). We note that u <u>is positive-definite and lies in the centre of</u> $L^1(G)$, on account of (8) and the corresponding properties of χ_{D_n}. Moreover, (10) and (13) show that

$$(14) \qquad \|u * f - f\|_S < \varepsilon \qquad f \in F.$$

We will now prove that u belongs to the ideal I, the function u being defined by (11). This will result from the following proposition:

$$(15) \qquad \text{If } f \in I \text{ and } \chi_D * f \neq 0, \text{ then } \chi_D \in I.$$

Here I is any closed two-sided ideal of a symmetric Segal algebra $S^1(G)$ and χ_D is a character of a finite-dimensional irreducible representation (of class) D of the compact group G.

To verify (15), we note that I, being a closed two-sided ideal of $S^1(G)$, is a two-sided ideal even in $L^1(G)$ (cf. § 8, Corollary of Proposition 1; here the symmetry of $S^1(G)$ is used). Now let I_D be the closed (necessarily two-sided) ideal generated by χ_D in $L^1(G)$. The condition $\chi_D * f \neq 0$ in (15) implies

$$(16) \qquad I_D \cap I \neq (0).$$

As is well-known (cf. also below), I_D has the following property:

(17) I_D contains no two-sided ideal other that (0) and I_D.

Thus, (16) implies $I_D \subset I$. As I_D contains χ_D (cf. (5)), we have proved (15). Hence u, defined by (11), lies in I, which concludes the proof of Theorem 1.

Let us now, for the sake of completeness, give a proof of (17). Write the representation (belonging to the equivalence class) D in matrix form

$$(18) \qquad x \to (m_{jk}(x)), \ 1 \leqslant j,k \leqslant \dim D,$$

and put

$$M_{jk}(x) = \dim D . m_{jk}(x) \qquad x \in G.$$

Then we have

$$(19) \qquad M_{jk} * M_{kl} = M_{jl}, \quad M_{jk} * M_{k'l} = 0 \text{ if } k' \neq k$$

(cf. WEIL [1], p. 73, where these relations are expressed in an equivalent way). Consider the linear subspace M_D of dimension $(\dim D)^2$ spanned by the functions M_{jk}. The formulae (19) show at once that M_D is an algebra under convolution, i.e. a subalgebra of $L^1(G)$, and that <u>the algebra</u> M_D <u>is isomorphic to the ordinary matrix algebra</u> $M(n,\mathbf{C})$, <u>with</u> $n = \dim D$. <u>Thus</u> M_D <u>is a simple algebra</u>, i.e. it contains no two-sided ideal other than (0) and M_D itself; the proof of this fact is both classical and very simple: if such an ideal contains a non-zero element, say

$$f = \Sigma_{j,k} a_{jk} M_{jk},$$

then it also contains, <u>for any fixed pair</u> j,k and $1 \leqslant p,q \leqslant n$, the element

$$M_{pj} * f * M_{kq} = a_{jk} M_{pq}$$

(cf. (19)); since here at least one $a_{jk} \neq 0$, the ideal contains M_{pq} for all $1 \leqslant p,q \leqslant n$, i.e. the ideal is all of M_D.

Next we observe that M_D <u>is a closed, two-sided ideal of</u> $L^1(G)$: indeed, M_D is a finite-dimensional linear subspace of $L^1(G)$, hence closed in $L^1(G)$; also, M_D is invariant under both left and right translations, since (cf. (18))

$$m_{jk}(ab) = \Sigma_r m_{jr}(a).m_{rk}(b) \qquad a,b \in G,$$

whence

$$M_{jk}(ax) = \Sigma_r m_{jr}(a).M_{rk}(x), \quad M_{jk}(xb) = \Sigma_r m_{rk}(b).M_{jr}(x).$$

The above implies that M_D contains I_D, the closed (two-sided) ideal generated in $L^1(G)$ by $\chi_D = \Sigma_j m_{jj}$, and as M_D is a simple algebra,

I_D <u>coincides with</u> M_D; thus (17) is proved. We observe that

$$(20) \qquad I_D = \{\chi_D * f \mid f \in L^1(G)\},$$

since χ_D lies in the centre of $L^1(G)$.

As we have just seen, the proof of Theorem 1 rests also on a fundamental fact of linear algebra.

COROLLARY 1. For a compact group G, a <u>symmetric</u> Segal algebra $S^1(G)$ contains <u>all ideals</u> I_D (cf. (20)), i.e. it contains I_D for every equivalence class D of finite-dimensional irreducible representations of G.

<u>Proof</u>. I_D is the closure of $I_D \cap S^1(G)$ in $L^1(G)$ (§ 9, Theorem 1), hence

$$I_D \cap S^1(G) \neq (0),$$

which implies, by (17),

$$I_D \subset S^1(G),$$

since $S^1(G)$ is a two-sided ideal in $L^1(G)$.

Corollary 1 is the analogue for compact groups of a result proved for abelian groups in Ch. 6, § 2.2 (iii).

COROLLARY 2. For a compact group G, every closed, two-sided ideal I of a <u>symmetric</u> Segal algebra $S^1(G)$ coincides with the closed linear subspace generated in $S^1(G)$ by all ideals I_D which I contains.

<u>Proof</u>. Note that, for any $f \in I$, the approximate units u of Theorem 1 belong to this subspace (cf. (11), (12), (15)), which is also a closed, two-sided ideal of $S^1(G)$; hence $u * f$ belongs to this subspace, and thus so does f.

For the case $S^1(G) = L^1(G)$, Corollary 2 is classical.

Let us now consider l.c. groups containing closed, normal abelian subgroups with compact quotient groups; here we can give the following applications of the results in § 14 and in the present section - applications analogous to those in § 14, (I) and (II), where soluble groups and abelian quotient groups were considered.

1. Let G be a l.c. group that contains closed, normal, abelian subgroups H_n with compact quotient groups; for instance, G may be a semidirect product of a compact group and a l.c. abelian group, or a product of finitely many such semidirect products. Examples of semidirect products are the motion groups of ν-dimensional euclidean space, $\nu \geqslant 2$. Consider a symmetric Segal algebra $S^1(G)$. Every closed, two-sided ideal J_n of $S^1(G)$ that contains $J^1_S(G,H_n)$ belongs to $\mathcal{J}_{ma}(S^1(G))$ and to $\mathcal{J}_{ma'}(S^1(G))$, and hence so does any finite intersection $\cap_n J_n$.

This follows from § 13, Theorem 1, and from Theorem 1 above, combined with § 14, Corollary 1. The analogy with the result given in § 14, (II_1) will be observed.

2. Let G and H_n be as in 1 and let, for each n, μ_n be a central, idempotent measure on G/H_n (on any compact group there clearly exist non-trivial measures of this kind, e.g. $d\mu(x) = (\Sigma_k \dim D_k . \chi_{D_k}).dx$, where $\int dx = 1$ and Σ_k is a finite sum). Then any finite intersection

$$\cap_n \chi_n . T^{-1}_{H_n}(\mu_n * L^1(G/H_n)),$$

where each χ_n is a character of G, belongs to $\mathcal{J}_{mb}(L^1(G))$ and to $\mathcal{J}_{mb'}(L^1(G))$.

This follows from § 14, Corollary 2: note that, since μ_n is central, $\mu_n * L^1(G/H_n)$ is in $\mathcal{J}_{mb}(L^1(G/H_n))$ and in $\mathcal{J}_{mb'}(L^1(G/H_n))$ (cf. § 14. (I_2) and (I_3); the multiple approximate units may here even be chosen to be central). This result is entirely analogous to § 14, (II_2).

3. Let G be a compact group. Let for each equivalence class D of finite-dimensional irreducible representations of G a number $a_D \geqslant 1$ be given; write $A = (a_D)$. Let $S^1_A(G)$ consist of all continuous functions

f on G such that

$$\|f\|_S = \Sigma_D \|\dim D.\chi_D * f\|_\infty . a_D$$

is finite; $S^1_A(G)$ is a symmetric Segal algebra, as is readily seen. Now let g be any continuous function on G such that $\chi_D * g \neq 0$ for infinitely many D. Then we put

$$a_D = \max(1, 1/\|\dim D.\chi_D * f\|_\infty) \text{ if } \chi_D * f \neq 0,$$

$$a_D = 1 \text{ if } \chi_D * f = 0.$$

Thus g does not belong to the corresponding Segal algebra $S^1_A(G)$. Combining this with Corollary 1 above, we see that for a compact group G the intersection of all symmetric Segal algebras $S^1(G)$ consists of the finite linear combinations of the matrix elements of inequivalent finite-dimensional irreducible representations of G. This is the analogue of § 5, (vii).

§ 16. Segal algebras on abelian groups

In § 9, Theorem 1, we established a bijective correspondence between the closed ideals of symmetric Segal algebras $S^1(G)$ and those of $L^1(G)$; if G is abelian, we can prove that this correspondence preserves the existence of approximate units.

THEOREM 1. Let G be any l.c. abelian group, $S^1(G)$ a Segal algebra on G. A closed ideal I_S of $S^1(G)$ belongs to $\mathcal{J}_a(S^1(G))$ if and only if the closure of I_S in $L^1(G)$ belongs to $\mathcal{J}_a(L^1(G))$ (for the notation, see § 14, (i)).

The proof is based on the method of linear functionals (§ 11) and requires some preparation; moreover, the assumption that G is abelian is needed only at one point of the proof. We shall, therefore, carry out our work for general l.c. groups, as long as possible, and introduce the commutativity of G only at the very end.

Let $S^1(G)$ be any Segal algebra, G being a general l.c. group. $S^1(G)$ acts in a natural way on the dual space $S^1(G)'$: for $f \in S^1(G)$ we define the operator $f^* \ast$ on $S^1(G)'$ by the relation (cf. § 11, (6) for the notation)

$$(1) \qquad \langle g\ ,\ f^* \ast \phi_S \rangle_S \ =\ \langle f \ast g\ ,\ \phi_S \rangle_S \qquad \text{for all } g \in S^1(G).$$

This means that the operator $f^* \ast$ is the adjoint of the left multiplication operator on $S^1(G)$; this is one reason for the notation. We note that

$$(2) \qquad (f_1 \ast f_2)^* \ast \phi_S \ =\ f_2^* \ast f_1^* \ast \phi_S .$$

Another reason is provided by the following lemma which will be essential for the proof of Theorem 1.

LEMMA 1. Let $S^1(G)$ be a <u>symmetric</u> Segal algebra. Then the functional $f^* \ast \phi_S$ in $S^1(G)'$, obtained from $\phi_S \in S^1(G)'$ by means of the operator $f^*\ast$ $(f \in S^1(G))$, corresponds to a bounded continuous function $x \to f^* \ast \phi_S(x)$ on G (abuse of notation!) and we have

$$(3) \qquad f^* \ast \phi_S(x) = \overline{\langle R_x f , \phi_S \rangle_S} \qquad x \in G.$$

In other words, if $f^* \ast \phi_S(x)$ is <u>defined</u> by (3), then we have

$$(4) \qquad \langle g , f^* \ast \phi_S \rangle_S = \int g(x)\overline{f^* \ast \phi_S(x)}\, dx \qquad g \in S^1(G).$$

In particular, if the functional ϕ_S corresponds to $\phi \in L^\infty(G)$, i.e. if ϕ_S is given by

$$f \to \int f(x)\overline{\phi(x)}\, dx \qquad f \in S^1(G),$$

then (3), with ϕ in the place of ϕ_S, agrees with the customary definition of the function $f^* \ast \phi$ as the convolution of f^* with ϕ.

<u>Proof</u>. The relation (4) has already been proved: it is another formulation of § 11, Lemma 1, relation (9), with f and g interchanged. The customary definition of the function $f^* \ast \phi$, for $\phi \in L^\infty(G)$, is

$$\int f^*(y)\phi(y^{-1}x)\, dy = \int \overline{f(y)}\phi(yx)\, dy,$$

which agrees with (3) when $\phi_S \in S^1(G)'$ corresponds to $\phi \in L^\infty(G)$ (cf. also § 11, (7)). Thus Lemma 1 is proved.

We note that (3) gives for $x = e$

$$(5) \qquad f^* \ast \phi_S(e) = \overline{\langle f , \phi_S \rangle_S}.$$

Another lemma useful in the proof of Theorem 1 is given below. We shall write $\phi_S \perp I$ ($\phi_S \in S^1(G)'$, $I \subset S^1(G)$) to indicate that $\langle f , \phi_S \rangle_S = 0$ for all $f \in I$.

LEMMA 2.

(i) Let $S^1(G)$ be <u>any</u> Segal algebra; let I be a <u>left</u> ideal (closed or not) of $S^1(G)$. Then $\phi_S \perp I$ holds if and only if $f^* * \phi_S \perp I$ holds for all $f \in S^1(G)$.

(ii) Let $S^1(G)$ be <u>symmetric or pseudosymmetric</u>; let I be a <u>right</u> ideal (closed or not) of $S^1(G)$. Then $\phi_S \perp I$ holds if and only if $f^* * \phi_S = 0$ for all $f \in I$.

<u>Proof</u>.

(i) If $\langle f_o , \phi_S \rangle_S = 0$ for each $f_o \in I$, then

$$\langle f * f_o , \phi_S \rangle_S = 0$$

for all $f \in S^1(G)$, since I is a left ideal, whence

$$\langle f_o , f^* * \phi_S \rangle_S = 0$$

for all $f \in S^1(G)$, and each $f_o \in I$. Conversely, if this last relation holds for each $f_o \in I$ and all $f \in S^1(G)$, then

$$\langle f * f_o , \phi_S \rangle_S = 0;$$

for fixed f_o we then take $f = u_n$, with $(u_n)_{n \geq 1}$ in $S^1(G)$ such that

$$u_n * f_o \to f_o \text{ in } S^1(G) \quad (n \to \infty),$$

which yields $\langle f_o , \phi_S \rangle_S = 0$ for all $f_o \in I$.

(ii) Suppose $\langle f , \phi_S \rangle_S = 0$ for all $f \in I$. Then, since I is a right ideal, we have

$$\langle f * g , \phi_S \rangle_S = 0 \quad \text{for all } g \in S^1(G),$$

whence

$$\langle g , f^* * \phi_S \rangle_S = 0 \quad \text{for all } g \in S^1(G),$$

that is

$$f^* * \phi_S = 0 \quad \text{for all } f \in I.$$

Conversely, if this last condition holds, then - working backwards -

we obtain

$$\langle f * g \, , \, \phi_S \rangle_S = 0 \quad \text{for every } g \in S^1(G).$$

Now $S^1(G)$, being symmetric or pseudosymmetric, has right approximate units (§ 8, Proposition 1, (ii), (iii)); thus we can, for fixed $f \in I$, put $g = u_n$, with $(u_n)_{n \geq 1}$ in $S^1(G)$ such that

$$f * u_n \to f \text{ in } S^1(G) \quad (n \to \infty).$$

This yields $\langle f \, , \, \phi_S \rangle_S = 0$ for all $f \in I$.

We can now prove some propositions from which Theorem 1 will follow.

PROPOSITION 1. Let $S^1(G)$ be a <u>symmetric</u> Segal algebra; let I_S be a closed <u>right</u> ideal of $S^1(G)$. Then I_S has approximate right units if and only if the following <u>condition</u> (C_S) is satisfied:

$$(C_S) \begin{cases} \text{Whenever } f \in I_S \text{ and } \phi_S \in S^1(G)' \text{ are such} \\ \text{that } f^* * \phi_S \perp I_S \text{ holds, then } f^* * \phi_S = 0. \end{cases}$$

<u>Proof</u>. Suppose I_S has approximate right units and let $f \in I_S$ and $\phi_S \in S^1(G)'$ be such that

$$f^* * \phi_S \perp I$$

holds. Choose $(u_n)_{n \geq 1}$ in I_S so that

$$f * u_n \to f \text{ in } S^1(G) \quad (n \to \infty);$$

then

$$u_n^* * (f^* * \phi_S) = 0 \quad \text{for each } n$$

(cf. Lemma 2 (ii)), or

$$(f * u_n)^* * \phi_S = 0$$

(cf. (2)), whence for $n \to \infty$ results:

$$f^* * \phi_S = 0.$$

Conversely, suppose (C_S) holds. Given any $f \in I_S$, consider the right ideal

$$I_f = \{f * g \mid g \in I_S\}$$

of $S^1(G)$; we want to show: f lies in the $S^1(G)$-closure of I_f. Consider any $\phi_S \perp I_f$: thus, by Lemma 2, (ii) again,

$$(f * g)^* * \phi_S = 0 \quad \text{for all } g \in I_S ,$$

or $g^* * (f^* * \phi_S) = 0$ for all $g \in I_S$, whence

$$f^* * \phi_S \perp I_S ,$$

by the same lemma. Thus, by the assumed condition (C_S), we have

$$f^* * \phi_S = 0 ,$$

whence $\langle f , \phi_S \rangle_S = 0$ (cf. Lemma 1 and relation (5)). Since $\phi_S \perp I_f$ was arbitrary, the desired result follows. Thus the proof is complete.

REMARK 1. In the case $S^1(G) = L^1(G)$, we simply write <u>condition</u> (C) instead of (C_S); explicitly this reads, for a closed <u>right</u> ideal I of $L^1(G)$:

$$\text{(C)} \quad \begin{cases} \text{Whenever } f \in I \text{ and } \phi \in L^\infty(G) \text{ are such} \\ \text{that } f^* * \phi \perp I \text{ holds, then } f^* * \phi = 0. \end{cases}$$

In this context we mention again that here $f^* * \phi$ may be interpreted in the customary way (cf. Lemma 1).

PROPOSITION 2. Let $S^1(G)$ be a <u>symmetric</u> Segal algebra and let I be a closed <u>two-sided</u> ideal of $L^1(G)$; put $I_S = I \cap S^1(G)$, so that I_S

is a closed, two-sided ideal of $S^1(G)$. If I_S satisfies (C_S), then I satisfies (C).

<u>Proof</u>. Let $f \in I$ and $\phi \in L^\infty$ be such that $f^* * \phi \perp I$. There is a sequence $(u_n)_{n\geq 1}$ in $S^1(G)$ such that

$$f * u_n \to f \text{ in } L^1(G) \quad (n \to \infty);$$

cf. part (i) of the proof of Theorem 1 in § 9. Now $f * u_n$ lies in $S^1(G)$ (since $S^1(G)$ is a left ideal in $L^1(G)$) and also in I (since I is, in particular, a right ideal); thus $f * u_n$ lies in I_S. Now we also have

$$u_n^* * (f^* * \phi) \perp I,$$

by Lemma 2, (i) (since I is also a left ideal), or

$$(f * u_n)^* * \phi \perp I,$$

and in particular

$$(f * u_n)^* * \phi \perp I_S.$$

Since $f * u_n \in I_S$, we can apply (C_S) which yields

$$(f * u_n)^* * \phi \;=\; 0,$$

and for $n \to \infty$ we obtain

$$f^* * \phi \;=\; 0,$$

i.e. (C) holds for I.

PROPOSITION 3. Let G be a l.c. <u>abelian</u> group and $S^1(G)$ a Segal algebra on G. Let I_S be a closed ideal of $S^1(G)$ and let I be the closure of I_S in $L^1(G)$. If I satisfies (C), then I_S satisfies (C_S).

<u>Proof</u>. Let $f \in I_S$ and $\phi_S \in S^1(G)'$ be such that

$$f^* * \phi_S \perp I_S.$$

There is a sequence $(u_n)_{n\geq 1}$ in $S^1(G)$ such that

$$u_n * f \to f \text{ in } S^1(G) \quad (n \to \infty).$$

Now we also have, for each n,

$$u_n^* * (f^* * \phi_S) \perp I_S,$$

by Lemma 2 (i); hence, <u>since</u> G <u>is abelian</u>,

$$f^* * (u_n^* * \phi_S) \perp I_S. \tag{6}$$

Note that here $u_n^* * \phi_S$ may be considered as a function ϕ_n in $L^\infty(G)$ (cf. Lemma 1), and $f^* * \phi_n$ is an ordinary convolution, by the same lemma, with f^* defined in the usual way. Passing now to I, the closure of I_S in $L^1(G)$, we obtain from (6)

$$f^* * (u_n^* * \phi_S) \perp I.$$

Since $f \in I_S \subset I$ and $u_n^* * \phi_S \in L^\infty(G)$, we can apply the assumed condition (C): it follows that, for each n,

$$f^* * (u_n^* * \phi_S) = 0,$$

or

$$(u_n * f)^* * \phi_S = 0.$$

Letting here $n \to \infty$, we obtain

$$f^* * \phi_S = 0,$$

i.e. (C_S) holds for I_S.

<u>Proof of Theorem</u> 1. This now results from § 9, Theorem 1 (or Ch. 6, § 2.4), and from Propositions 1, 2, 3 above.

Theorem 1 reduces the investigation of $\mathcal{J}_a(S^1(G))$, for <u>abelian</u> G, to that of $\mathcal{J}_a(L^1(G))$; incidentally, already for abelian G <u>not all</u>

closed ideals of $L^1(G)$ belong to $\mathcal{J}_a(L^1(G))$ if G is not compact (cf. § 17), in contrast to § 15, Theorem 1.

It is an open question whether for Segal algebras on abelian l.c. groups G every ideal in $\mathcal{J}_a(S^1(G))$ possesses approximate units having a positive Fourier transform with compact support: this would be some analogy to § 15, Theorem 1. In the other direction, it is also an open question how far Theorem 1 above can be extended to general l.c. groups.

In connection with Theorem 1 let us restate a familiar definition.

DEFINITION. Let G be a l.c. abelian group, $S^1(G)$ a Segal algebra. A closed set $\hat{E}$ in the dual group $\hat{G}$ is said to be a Wiener-Ditkin set for $S^1(G)$ if there is only one closed ideal I_S of $S^1(G)$ such that cosp $I_S = \hat{E}$ (i.e. if $\hat{E}$ is a Wiener set for $S^1(G)$) and if this ideal I_S belongs to $\mathcal{J}_a(S^1(G))$.

This definition agrees with that given in Ch. 2, § 5.2 (cf. especially the Remark loc. cit.).

We now have immediately the following result:

COROLLARY of Theorem 1. For any Segal algebra $S^1(G)$ the Wiener-Ditkin sets are the same as those for $L^1(G)$.

Proof. This is a simple consequence of § 9, Theorem 1 (or Ch. 6, § 2.4), and Theorem 1 above.

In particular, single points of the dual group $\hat{G}$ are Wiener-Ditkin sets for $S^1(G)$. It is useful to verify directly that a Segal algebra satisfies the condition of Wiener-Ditkin: given any $\hat{a} \in \hat{G}$ and $f \in S^1(G)$ such that $\hat{f}(\hat{a}) = 0$, there is for every $\varepsilon > 0$ a $\tau \in S^1(G)$ such that

$$\hat{\tau}(\hat{x}) = 1 \text{ near } \hat{a} \text{ and } \|f * \tau\|_S < \varepsilon.$$

This can be verified very simply by reduction to $L^1(G)$, as follows. There is an $h \in S^1(G)$ such that $\hat{h}$ has compact support and $\hat{h}(\hat{x}) = 1$ near $\hat{a}$ (cf. Ch. 6, § 2.2 (iii)). $L^1(G)$ satisfies the condition of Wiener-Ditkin (cf. Ch. 6, § 1.4), thus there is a $\tau_1 \in L^1(G)$ such that $\hat{\tau}_1(\hat{x}) = 1$ near $\hat{a}$ and

$$\|f * \tau_1\|_1 < \varepsilon/\|h\|_S.$$

Let us put $\tau = \tau_1 * h$; then $\tau \in S^1(G)$ (since $h \in S^1(G)$), $\hat{\tau}(\hat{x}) = 1$ near $\hat{a}$, and

$$\|f * \tau\|_S \leqslant \|f * \tau_1\|_1 . \|h\|_S < \varepsilon.$$

REMARK 2. By the Corollary above, we may now simply speak of ' Wiener-Ditkin sets in $\hat{G}$ ', without any ambiguity. Likewise we may say ' Wiener set in $\hat{G}$ ': this is already familiar from Ch. 6, § 2.4. In practice, we are thus free to consider only $L^1(G)$.

Let us finally discuss anew the injection theorem for Wiener-Ditkin sets which was given in Ch. 7, § 4.5 and reads as follows. Let Γ be a closed subgroup of the dual group $\hat{G}$ and let $\hat{E}$ be a closed subset of Γ. Then $\hat{E}$ is a Wiener-Ditkin set in Γ if and only if $\hat{E}$ is a Wiener-Ditkin set in $\hat{G}$.

With the tools at our disposal, we can give the proof as follows. First, $\hat{E}$ is a Wiener set in Γ if and only if $\hat{E}$ is a Wiener set in $\hat{G}$: this is the injection theorem for Wiener sets (cf. Ch. 7, § 3.8). Now let $H \subset G$ be the orthogonal subgroup of $\Gamma \subset \hat{G}$, i.e. $(G/H)^\wedge = \Gamma$. Let I be the (unique) closed ideal of $L^1(G/H)$ with cospectrum $\hat{E}$; then the (unique) closed ideal of $L^1(G)$ with cospectrum $\hat{E}$ is clearly $T_H^{-1}(I)$ (cf. Ch. 4, § 4.3). If I lies in $\mathcal{J}_a(L^1(G/H))$, then $T_H^{-1}(I)$ lies in $\mathcal{J}_a(L^1(G))$: see § 14, Corollary 1 (with $S^1(G) = L^1(G)$ and $n = 1$); the converse is obvious, which completes the proof.

REMARK 3. The proof of the injection theorem for Wiener-Ditkin sets given here should be compared with that in Ch. 7, §§ 4.4 and 4.7. Corollary 1 of § 14 (cf. also Lemma 2 of § 8) represents, as it were, the 'non-commutative part' of that proof. A comparison of the two proofs will show that by separating the non-commutative and the strictly commutative parts one obtains a simpler proof and a clearer insight into the structure. Similarly, the conditions (C_S) and (C) used in the proof of Theorem 1 above embody the non-commutative part of the criterion of Herz-Glicksberg (Ch. 7, § 4.9).

In connection with Remark 3 the following may be mentioned. It often occurs that the proof of a result in classical harmonic analysis may be divided into two parts: one that admits of an extension to non-abelian groups and another, strictly abelian one. A clear recognition of these two components is of considerable interest: it leads not only to more general results, but also to greater simplicity in the proofs.

§ 17. L^1-algebras of compact and of abelian groups

We shall now study $\mathcal{J}_b(L^1(G))$ for compact and for locally compact abelian groups. Let us first consider the compact case.

THEOREM 1. Let G be a compact group and let I be a closed, two-sided ideal of $L^1(G)$. If I has bounded approximate left units, then I is of the form

$$I = \mu * L^1(G), \tag{1}$$

where μ is an idempotent and central measure in $M^1(G)$. Moreover, every ideal of the form (1), with μ as stated, is closed and two-sided and has bounded multiple approximate two-sided units u of the form

$$u = \Sigma_n\, c_n . \chi_{D_n} \text{ with } c_n > 0, \tag{2}$$

where the sum is finite; thus u is positive-definite, in particular self-adjoint, and lies in the centre of $L^1(G)$. Moreover, u can also be so chosen that in addition

$$\|u\|_1 \leq \|\mu\|. \tag{3}$$

REMARK 1. Theorem 1 contains in particular the assertion that

$$\mathcal{J}_b(L^1(G)) = \mathcal{J}_{mb}(L^1(G)), \tag{4}$$

if G is compact. We also note that an ideal I of the type (1), with μ as above, is stable under involution.

Proof. Let $I \in \mathcal{J}_b(L^1(G))$ be given. Consider any finite set F of characters χ_D belonging to I; put

$$f = \Sigma_{\chi_D \in F} \dim D . \chi_D,$$

so that $f \in I$. Given $\varepsilon > 0$, there is by hypothesis a $u \in I$ such that

(5) $$\|u * f - f\|_1 < \varepsilon$$

and

(6) $$\|u\|_1 \leq C,$$

where C <u>is independent of</u> F <u>and</u> ε. We have

$$f * \chi_D = \chi_D \qquad \chi_D \in F,$$

hence we obtain from (5)

(7) $$\|u * \chi_D - \chi_D\|_1 < \varepsilon \qquad \chi_D \in F,$$

since $\|\chi_D\|_1 \leq \|\chi_D\|_2 = 1$ (for the properties of characters used here cf. WEIL [1], p. 78).

Denote by $A(F,\varepsilon)$ the set of all $u \in I$ satisfying (6) and (7); we may embed $A(F,\varepsilon)$ in $M^1(G)$ in the standard way. We have shown that $A(F,\varepsilon)$ is not empty. Let $B(F,\varepsilon)$ be the closure of $A(F,\varepsilon)$ in the $\sigma(M^1(G), \mathcal{K}(G))$-topology. In this topology the set $B(F,\varepsilon)$ is compact, being a closed subset of the compact ball of radius C and centre at the origin (cf. (6)). Moreover, the family $(B(F,\varepsilon))$ (F any finite set of characters χ_D in I, and $\varepsilon > 0$) has the finite intersection property, as this is true for the corresponding family $(A(F,\varepsilon))$. Thus the intersection

$$\bigcap_{F,\varepsilon} B(F,\varepsilon) \quad (F \text{ any finite set of characters } \chi_D \in I,\ \varepsilon > 0)$$

is non-empty; let μ be a measure contained in this intersection. We shall show: the measure μ satisfies

(8a) $\mu * \chi_D = \chi_D$ if $\chi_D \in I$; (8b) $\mu * \chi_D = 0$ if $\chi_D \notin I$.

To verify (8a) observe that any $u \in A(F,\varepsilon)$ satisfies, in virtue of (7), for all $\chi_D \in F$

$$\left| \int u * \chi_D(x)f(x)dx - \int \chi(x)f(x)dx \right| < \varepsilon.\|f\|_\infty \qquad f \in \mathcal{K}(G).$$

As

$$\int u * \chi_D(x)f(x)dx = \int\left\{\int \chi_D(y^{-1}x)f(x)dx\right\}u(y)dy,$$

it follows that <u>any</u> $\mu' \in B(F,\varepsilon)$ <u>satisfies for all</u> $\chi_D \in F$

$$\left| \int\left\{\int \chi_D(y^{-1}x)f(x)dx\right\}d\mu'(y) - \int \chi_D(x)f(x)dx \right| \leqslant \varepsilon.\|f\|_\infty,$$

by continuity in the topology $\sigma(M^1(G), \mathcal{K}(G))$; that is, we have

$$\left| \int \mu' * \chi_D(x)f(x)dx - \int \chi_D(x)f(x)dx \right| \leqslant \varepsilon.\|f\|_\infty \qquad f \in \mathcal{K}(G),$$

for $\mu' \in B(F,\varepsilon)$ and $\chi_D \in F$. Hence a measure μ contained in <u>all</u> $B(F,\varepsilon)$ satisfies (8a). To prove (8b), we observe that, if $\chi_D \notin I$, then any $u \in I$ satisfies (cf. § 15, (15))

$$\int \chi_D(y^{-1}x)u(y)\,dy = 0 \qquad \text{for each } x \in G,$$

and it follows, by continuity, that (8b) holds. The measure $\mu \in M^1(G)$ is of course uniquely determined by (8a,b).

Let us write (8a.b) in the (somewhat less precise) form

(9) $$\mu * \chi_D = \lambda_D.\chi_D, \quad \lambda_D = 0 \text{ or } 1 \text{ for every } D.$$

(9) <u>implies that the measure</u> μ <u>is idempotent and central</u>. To show this, consider any $f \in K(G)$; the function $\phi = \mu * f$ has the 'Fourier series' (cf. WEIL [1], p. 77).

(10) $$\Sigma_D \dim D.\lambda_D.\chi_D * f,$$

by (9), since

$$\chi_D * (\mu * f) = (\mu * f) * \chi_D = (\mu * \chi_D) * f.$$

The Fourier series of $\mu * \phi$ coincides with (10), which implies $\mu * \phi = \phi$ or, as $f \in \mathcal{K}(G)$ was arbitrary, $\mu * \mu = \mu$. Moreover, we have

$$(11) \qquad \mu * f = f * \mu \qquad f \in \mathcal{K}(G),$$

since both sides have the same Fourier series (10). This implies $\mu * \nu = \nu * \mu$ for all $\nu \in M^1(G)$: for, replacing in (11) f by $\nu * f$, we obtain

$$\mu *(\nu * f) = (\nu * f) * \mu = \nu *(f * \mu) = \nu *(\mu * f)$$

or $(\mu * \nu) * f = (\nu * \mu) * f$ for all $f \in \mathcal{K}(G)$, whence $\mu * \nu = \nu * \mu$. We observe that <u>the condition</u> (9) <u>is also necessary for a measure</u> μ <u>to be idempotent and central</u>. Indeed, let μ be such a measure and consider $\mu * \chi_D$, for any D. Then $\mu * \chi_D$ is central in $L^1(G)$; moreover, by § 15, (5) we have

$$\mu * \chi_D = \dim D.(\mu * \chi_D) * \chi_D,$$

so that $\mu * \chi_D$ lies in the (closed, two-sided) ideal generated in $L^1(G)$ by χ_D. The centre of the algebra $I_D \cong M(n,\mathbb{C})$ (n = dim D) (cf. § 15, (19), (20) and the connecting text) consists of the scalar multiples of χ_D, by Schur's lemma, as is well-known; thus, since $\mu * \chi_D$ is in particular central in I_D, we have

$$\mu * \chi_D = \lambda_D . \chi_D,$$

with λ_D a scalar, and since μ is idempotent, it follows that $\lambda_D = 1$ or 0.

Let us now return to the particular measure μ satisfying (8a,b). We want to show that, with this μ, the relation (1) holds. We recall § 15, Corollary 2, for the classical case $S^1(G) = L^1(G)$, and the fact that $\mu * L^1(G)$ is a <u>closed</u> (two-sided) ideal of $L^1(G)$ if μ is idempotent (and central), as shown in § 14, (I_2); moreover,

$$\text{if } \chi_D \in I, \text{ then } \chi_D \in \mu * L^1(G),$$

by (8a), and

$$\text{if } \chi_D \notin I, \text{ then } \chi_D \notin \mu * L^1(G),$$

since otherwise we would clearly have $\mu * \chi_D = \chi_D$, in contradiction to (8b). Thus (1) is proved.

Now let, conversely, an ideal I of the form (1) be given, with $\mu \in M^1(G)$ idempotent and central. $L^1(G)$ possesses multiple approximate two-sided units which are finite linear combinations of characters χ_D with positive coefficients <u>and have</u> L^1-<u>norm</u> 1; this follows from the initial part of the proof of Theorem 1 in § 15, with $I = S^1(G) = L^1(G)$: indeed, given any finite set $F \subset L^1(G)$ and $\varepsilon > 0$, we can take $w \in L^1(G)$ as in § 15, (7), (8), so that § 15, (10) holds; then we put

$$w' = w/\|w\|_1 .$$

In this way we obtain

$$\begin{aligned} \|w' * f - f\|_1 &\leqslant \|w * f - f\|_1 + \|w' - w\|_1 . \|f\|_1 \qquad f \in F \\ &< \varepsilon + |1 - \|w\|_1| . \|f\|_1 \\ &< (3/2).\varepsilon, \end{aligned}$$

the last step following from § 15, (9), since $1 = \int v * v^* = \|v * v^*\|_1$. [The same method will show that, for symmetric Segal algebras $S^1(G)$ on a compact group G, an analogue of § 8, Proposition 2 holds.] Now we put

$$u = \mu * w',$$

with μ as in (1) and w' as above. Recalling that w' is central, we see that the functions $u \in I$ may be used as multiple approximate <u>two-sided</u> units in I (cf. § 14, (6), (7), for $S^1(G) = L^1(G)$) and that they have the properties stated in Theorem 1, so that the proof of Theorem 1 is now complete.

We pass next to the consideration of the abelian case; here the following analogue of Theorem 1 holds:

THEOREM 2. Let G be a locally compact abelian group and I a closed ideal of $L^1(G)$. If I has <u>bounded</u> approximate units, then I is of the form

$$I = \bigcap_{1 \leq n \leq N} \chi_n \cdot T_{H_n}^{-1}(\mu_n * L^1(G/H_n)); \tag{12}$$

here N is a strictly positive integer and, for each n, χ_n is a character of G, H_n is a closed subgroup of G, and μ_n is an <u>idempotent</u> measure in $M^1(G/H_n)$. Moreover, an ideal of the form (12) is closed and has bounded <u>multiple</u> approximate units having a <u>positive</u> Fourier transform with <u>compact</u> support; thus these approximate units are positive-definite, in particular self-adjoint.

REMARK 2. Theorem 2 contains in particular the assertion that (4) also holds for l.c. abelian G. We observe that every ideal of the form (12) is stable under involution.

<u>Proof of Theorem</u> 2.

<u>First part of the proof</u>. Let an ideal $I \in \mathfrak{I}_b(L^1(G))$ be given; we will prove in steps (i)-(vi) below that I is of the form (12).

(i) Let $\hat{E} \subset \hat{G}$ be the cospectrum of I, i.e. the set of all $\hat{a} \in \hat{G}$ such that $\hat{f}(\hat{a}) = 0$ for <u>all</u> $f \in I$. Let $\overline{G}$ be the Bohr compactification of G; we recall that the dual of $\overline{G}$ is $\hat{G}_d$, the group $\hat{G}$ with the discrete topology (WEIL [1], § 35). We show that <u>there exists an idempotent measure</u> $\mu \in M^1(\overline{G})$ <u>such that the Fourier transform</u> $\hat{\mu}$ <u>coincides with the characteristic function of</u> $\hat{G}-\hat{E}$.

The <u>proof</u> is as follows. Given any finite subset $\hat{F} \subset \hat{G}-\hat{E}$, there is an $f \in I$ such that $\hat{f}(\hat{x}) = 1$ for all $\hat{x} \in \hat{F}$ (cf. Ch. 2, § 1.3, Remark, Ch. 2, § 1.4 (ii), and Ch. 6, § 1.3). Now, by assumption, I has bounded approximate units, thus for any $\varepsilon > 0$ there is a $u \in I$ such that

$$\|u * f - f\|_1 < \varepsilon \tag{13}$$

and

(14) $$\|u\|_1 \leqslant C,$$

where C is a constant independent of $f \in I$ and of $\varepsilon > 0$. Now let $A(\hat{F},\varepsilon)$ be the set of all $u \in I$ which satisfy (14) and are such that

(15) $$|\hat{u}(\hat{x}) - 1| < \varepsilon \qquad \text{for all } \hat{x} \in \hat{F}.$$

By the preceding, $A(\hat{F},\varepsilon)$ is non-empty, for any finite subset $\hat{F}$ of $\hat{G}-\hat{E}$ and any $\varepsilon > 0$: indeed, (13) implies (15). Now let G be embedded in $\overline{G}$, in the canonical way. Then we can embed $A(\hat{F},\varepsilon)$ in $M^1(\overline{G})$ by making correspond to the function u the measure $u(x)dx$, considered as a measure on $\overline{G}$; this embedding is isometric, i.e. we have

(16) $$\sup_\psi \left| \int_G \psi(x)u(x)\,dx \right| = \|u\|_1 \qquad \|\psi\|_\infty \leqslant 1,$$

where ψ ranges over the (continuous) almost-periodic functions on G. [The validity of (16) for any $u \in K(G)$ follows from the fact that, given any compact set $K \subset G$ and any continuous function k on G, there is clearly an almost-periodic function ψ on G coinciding with k on K; then the general validity of (16) for $u \in L^1(G)$ results by continuity.] Let $B(\hat{F},\varepsilon)$ be the closure of $A(\hat{F},\varepsilon)$ in $M^1(\overline{G})$ in the topology $\sigma(M^1(\overline{G}),\mathcal{K}(\overline{G}))$. Then we have by continuity (in this topology): <u>every</u> $\mu \in B(\hat{F},\varepsilon)$ <u>satisfies</u> (cf. (15))

(17a) $$|\hat{\mu}(\hat{x}) - 1| \leqslant \varepsilon \quad \text{for all } \hat{x} \in \hat{F} \subset \hat{G}-\hat{E}$$

Moreover, we also have by continuity

(17b) $$\hat{\mu}(\hat{x}) = 0 \quad \text{for all } \hat{x} \in \hat{E},$$

since for all $u \in I$ - in particular for $u \in A(\hat{F},\varepsilon)$ - we have: $\hat{u}(\hat{x}) = 0$ for each $\hat{x} \in \hat{E}$.

The set $B(\hat{F},\varepsilon)$ is compact, in the topology considered, for it is a closed subset of the compact ball of radius C around the origin in

$M^1(\overline{G})$ (cf. (14)). Further, the family $(B(\hat{F},\varepsilon))$ has the finite intersection property, since this is obviously true of the family $(A(\hat{F},\varepsilon))$; hence the intersection

$$\cap_{\hat{F},\varepsilon} B(\hat{F},\varepsilon) \quad (\hat{F} \subset \hat{G}-\hat{E} \text{ finite}, \ \varepsilon > 0)$$

is not empty. A measure μ contained in <u>all</u> sets $B(\hat{F},\varepsilon)$ satisfies in virtue of (17a,b)

$$\hat{\mu}(\hat{x}) = 1 \text{ if } \hat{x} \in \hat{G}-\hat{E}, \ \hat{\mu}(\hat{x}) = 0 \text{ if } \hat{x} \in \hat{E}.$$

Thus μ has the required property (and is, of course, unique).

(ii) It now follows from a well-known theorem of P.J. Cohen (for a simple proof of which we refer to ITÔ-AMEMIYA [1]) that $\hat{E}$ = cosp I <u>belongs to the coset ring of</u> $\hat{G}_d$ or, what amounts to the same, of $\hat{G}$:

$$\hat{E} \in \mathcal{R}(\hat{G}),$$

where $\mathcal{R}(\hat{G})$ denotes the Boolean algebra generated in $\hat{G}$ by <u>all</u> subgroups (closed or not) of $\hat{G}$ and their cosets.

(iii) Since cosp I is a <u>closed</u> subset of $\hat{G}$, the problem now arises how the elements of $\mathcal{R}(\hat{G})$ that are closed in $\hat{G}$ can be characterized. The solution is as follows: a set $\hat{E} \in \mathcal{R}(\hat{G})$ is closed if and only if it is of the form

$$\hat{E} = \cup_{1 \leq n \leq N} \lambda_n . \hat{E}_n, \tag{18}$$

where N is a strictly positive integer, λ_n is an element of $\hat{G}$, and

$$\hat{E}_n \in \mathcal{R}_o(\Gamma_n) \qquad 1 \leq n \leq N;$$

here Γ_n is a closed subgroup of $\hat{G}$ and $\mathcal{R}_o(\Gamma_n)$ denotes the <u>open</u> coset ring of Γ_n, i.e. the Boolean algebra generated in Γ_n by the (relatively) open subgroups of Γ_n and their cosets (in Γ_n). This result was established by GILBERT [1], Theorem 3.1, and later, but independently and with a far simpler proof, by SCHREIBER [1], Theorem 1.7; it applies even to arbitrary topological abelian groups $\hat{G}$.

(iv) Each set $\hat{E}_n$ in (18) is a Wiener-Ditkin set in $\hat{G}$. First, each $\hat{E}_n$ in (18) is a Wiener set in Γ_n; this is quite trivial, as $\hat{E}_n$ is both closed and open in Γ_n. Now let H_n be the closed subgroup of G orthogonal to Γ_n so that

$$(G/H_n)^{\wedge} = \Gamma_n.$$

As $\hat{E}_n$ lies in $\mathcal{S}_o(\Gamma_n)$, there is an idempotent measure μ_n in $M^1(G/H_n)$ such that $\hat{\mu}_n$ is the characteristic function of $\Gamma_n - \hat{E}_n$, by (the trivial part of) Cohen's theorem mentioned in (ii). Thus $\mu_n * L^1(G/H_n)$ is the only closed ideal of $L^1(G/H_n)$ with cospectrum $\hat{E}_n$; moreover, $\mu_n * L^1(G/H_n)$ has (bounded, multiple) approximate units (cf. § 14, (I_3)). Hence $\hat{E}_n$ is a Wiener-Ditkin set in Γ_n. The injection theorem for Wiener-Ditkin sets (cf. the final part of § 16) then shows that $\hat{E}_n$ is a Wiener-Ditkin set in $\hat{G}$, and hence so is the 'translated' set $\lambda_n . \hat{E}_n$.

(v) It follows from (iv) that the set $\hat{E}$ in (18) is a Wiener-Ditkin set in $\hat{G}$ (cf. Ch. 2, § 5.3 (iii)); in particular, $\hat{E}$ is a Wiener set in $\hat{G}$ (GILBERT [1], Theorem 3.9).

(vi) From (v) we now obtain: the given ideal I satisfies (12), where the idempotent measures μ_n in $M^1(G/H_n)$ are those introduced in (iv), and the characters χ_n of G are determined by the elements $\lambda_n \in \hat{G}$: in the usual notation

$$\chi_n(x) = \langle x, \lambda_n \rangle \qquad x \in G.$$

Indeed, with this choice of μ_n and χ_n, both sides of (12) have the same cospectrum (18).

This finishes the first part of the proof of Theorem 2. It will be observed that we had to resort to the use of Wiener-Ditkin sets in order to prove that the set $\hat{E}$ in (18) is a Wiener set (cf. (iv), (v)).

Second part of the proof. Let an ideal I of the form (12), with each μ_n idempotent, be given. That I has bounded multiple approximate units is already contained in § 14, (II_3), where also non-abelian groups were considered. It is very easy to deduce that, for abelian G, these approximate units can also be so chosen that they have the properties stated in Theorem 1; this we now proceed to show.

First we observe that an ideal of the form (12) is stable under involution: this is so for each ideal $\mu_n * L^1(G/H_n)$, for μ_n satisfies $\mu_n^* = \mu_n$, as both sides have <u>here</u> the same Fourier transform, and we have (as the groups are abelian)

$$(\mu_n * g)^* = \mu_n^* * g^* \qquad g \in L^1(G/H_n)$$

(for the definition of μ_n^* cf. Ch. 3, § 5.4, (14); there is an obvious misprint loc. cit., p. 77, line 2). Then it follows from § 1, (20) that I itself is stable under involution.

Now let any finite set $F \subset I$ and $\varepsilon > 0$ be given. Then we consider the functions f <u>and</u> f* in I, with $f \in F$, i.e. we replace F by

$$F \cup F^* \subset I.$$

As mentioned above, it has already been proved that I has <u>bounded multiple</u> approximate units, thus we can find a $w \in I$ such that

(19) $\|w * f - f\|_1 < \varepsilon$ <u>and</u> $\|w * f^* - f^*\|_1 < \varepsilon$ for all $f \in F$

and

(20) $$\|w\|_1 \leqslant C,$$

where C <u>is independent of</u> F <u>and</u> ε. The second inequality in (19) gives by involution

$$\|w^* * f - f\|_1 < \varepsilon \qquad f \in F,$$

as G is abelian, and this yields, in view of (20),

$$\|w * w^* * f - w * f\|_1 < C.\varepsilon \qquad f \in F.$$

Combining this with the first inequality in (19) we obtain, on putting

(21) $$w_o = w * w^*,$$

that

(22) $$\|w_o * f - f\|_1 < (C + 1).\varepsilon \qquad f \in F.$$

Also, w_o <u>still lies in</u> I and statisfies (cf. (20))

(23) $$\|w_o\|_1 \leqslant C^2.$$

Moreover, w_o obviously has a <u>positive</u> Fourier transform. Now we can choose $v_o \in L^1(G)$ such that

(24) $$\|v_o * w_o - w_o\|_1 < \varepsilon/M,$$

with $M = \max \|f\|_1$ $(f \in F)$, and also such that

(25) $$\|v_o\|_1 = 1$$

and

(26) the F.t. $\hat{v}_o$ is <u>positive</u> and has <u>compact</u> support

(cf. § 8, Proposition 2, for $S^1(G) = L^1(G)$). Then we obtain from (22) and (24), on putting

$$u = v_o * w_o,$$

that

$$\|u * f - f\|_1 < (C + 2).\varepsilon \qquad f \in F$$

holds. Moreover, u has a <u>positive</u> Fourier transform (since this holds for v_o and w_o) with <u>compact</u> support (cf. (26)). Also, u <u>lies in</u> I, since w_o does, and by (23) and (25) we have

$$\|u\|_1 \leqslant C^2.$$

As C is independent of F and ε, the (second part of the) proof of Theorem 2 is complete.

REFERENCES for Theorems 1 and 2. In Theorem 1 two-sided ideals are considered; an analogous result holds for one-sided ideals (even in the case of l.c. groups having the property P_1): see LIU-van ROOIJ-WANG [1], Theorem 4. As to Theorem 2, compare Theorem A in GILBERT [1]; the main part of Theorem 2 was proved independently in LIU-ROOIJ-WANG [2], Theorem 3.3. This last paper also contains a proof of the theorem of Gilbert-Schreiber which is one of the essential tools in the proof of Theorem 2: cf. part 1, (ii) of the proof above; for parts 1, (i) and 1, (ii) compare the proof of Theorem 1.1 in ROSENTHAL [1]. Finally we refer again to SCHREIBER [1] for the results and references to be found there.

REMARK 3. In connection with the last part of Theorem 2 we refer to § 12 (viii): the general method of proof of Theorem 2 above would only yield the bound 4 in the special case considered there.

To terminate let us discuss some matters concerning the families $\mathcal{J}_a(L^1(G))$ and $\mathcal{J}_{ma}(L^1(G))$, for G l.c. abelian.

1. If G is not compact, then $L^1(G)$ contains a closed ideal that does not belong to $\mathcal{J}_a(L^1(G))$ (this contrasts with compact groups, cf. § 15, Theorem 1).

Proof. There exists a function $f \in L^1(G)$ such that f is not contained in the closed ideal generated in $L^1(G)$ by $f * f$ (cf. below). Then (f), the closed ideal generated by f in $L^1(G)$, does not possess approximate units. Indeed, consider the function f which belongs to (f): the closure of

$$\{u * f \mid u \in (f)\}$$

does not contain f, for this closure coincides with the closure of

$$\{g * f * f \mid g \in L^1(G)\},$$

i.e. with the closed ideal generated by $f * f$. For infinite discrete G the existence of an $f \in L^1(G)$ with the property used above is contained in the profound work of MALLIAVIN [1]; for $\mathbb{R}^\nu$, $\nu \geq 3$, the proof is much easier (REITER [1]). For $\nu = 1$ one can readily deduce the existence of a function $f \in L^1(G)$ with the required property from MALLIAVIN's result for $L^1(\mathbb{Z})$, as obligingly shown to the author by Dr. J.D. Stegeman (however, a <u>direct</u> proof for $L^1(\mathbb{R})$ is still lacking). To pass from here to general l.c. (but not compact) abelian groups has now become a routine matter: such a group has a quotient group that is either $\mathbb{R}$ or an infinite discrete group, by a well-known structure theorem (WEIL [1], p. 110), and we may then apply a familiar fact (Ch. 3, § 5.3).

We note that this result extends, of course, to <u>any</u> l.c. group G such that G/G' is not compact, where G' is the (closure of) the commutator subgroup of G.

2. Are the families $\mathcal{J}_a(L^1(G))$ and $\mathcal{J}_{ma}(L^1(G))$ distinct, for non-compact G? An analogous problem may, of course, be raised for (commutative) Banach algebras.

3. If $I \in \mathcal{J}_a(L^1(G))$ is stable under involution, does I necessarily have approximate units which are self-adjoint, if G is not compact? We note that, if here $\mathcal{J}_a$ is replaced by $\mathcal{J}_{ma}$, then the answer is clearly affirmative, since

$$\|u * f - f\|_1 < \varepsilon \quad \text{and} \quad \|u * f^* - f^*\|_1 < \varepsilon$$

imply (as G is abelian)

$$\|\tfrac{1}{2}(u + u^*) * f - f\|_1 < \varepsilon$$

(for general compact groups cf. § 15, Theorem 1; cf. also Theorems 1 and 2 above).

4. If G is not compact, is there an ideal $I \in \mathcal{J}_a(L^1(G))$ such

that cosp I is <u>not</u> a Wiener set in $\hat{G}$?

5. Is there, for non-compact G, a Wiener set in $\hat{G}$ such that the corresponding closed ideal of $L^1(G)$ does <u>not</u> belong to $\mathcal{J}_a(L^1(G))$, in other words, are there Wiener sets in $\hat{G}$ that are <u>not</u> Wiener-Ditkin sets?

With this question we return to a problem that was raised in Ch. 7, § 4.10 and is still unsolved; it forms the end of these lectures.

§ 18. Appendix

The following changes should be made in the author's monograph 'Classical harmonic analysis and locally compact groups' (Oxford University Press, 1968).

P. ix (Reader's guide), footnote. REPLACE: des BY: de

P. 7, line 8 from below. REPLACE: τ_1' BY: $\hat{\tau}_1'$

P.16, line 13 from below. ADD AT THE END: , where w is subject to 6.1 (i)-(v)

P. 17, line 8. ADD AT THE END: , for the weight functions considered here.

P. 20, line 3. REPLACE: denoted cosp I BY: denoted by cosp I

P. 21, line 5. ADD: In the examples to be considered in this book, the underlying topological vector space of $\mathbf{A}(X)$ will, in fact, always be locally convex.

P. 29, § 4.8. A partial answer to the question raised here has been given by T.W. Körner, Proc. Cambridge Phil. Soc. 67 (1970), 559-568; see also the review in Zentralblatt f. Math. 194 (1970), 446.

P. 30, lines 8-10 from below. The proof as given applies only to the case where the underlying topological vector space of $\mathbf{A}(X)$ is locally convex, i.e.

$$\tfrac{1}{2}\,\mathcal{U}_0 + \tfrac{1}{2}\,\mathcal{U}_0 \subset \mathcal{U}_0 .$$

But the argument can easily be made perfectly general: we simply replace $\frac{1}{2}\mathcal{U}_0$ by a symmetric neighbourhood $\mathcal{U}_0'$ such that

$$\mathcal{U}_0' + \mathcal{U}_0' \subset \mathcal{U}_0 .$$

if $\mathcal{U}_0$ is any pre-assigned neighbourhood of 0 in $\mathbf{A}(X)$

P. 32, lines 10-16. Here the neighbourhood $\frac{1}{2}\mathcal{U}_0$ is used, which presupposes that $\mathcal{U}_0$ is convex (cf. the correction to p. 30). To make the reasoning valid in the general case, replace $\frac{1}{2}\mathcal{U}_0$ by $\mathcal{U}_0'$, where $\mathcal{U}_0'$ is so chosen that

$$\mathcal{U}_0' + \mathcal{U}_0' \subset \mathcal{U}_0,$$

and then replace $(2N)^{-1}\mathcal{U}_0$ by $\mathcal{U}_0''$, choosing $\mathcal{U}_0''$ so that

$$\mathcal{U}_0'' + \dots + \mathcal{U}_0'' \quad \text{(N times)}$$

is contained in $\mathcal{U}_0'$. Likewise for <u>lines 23 and 25</u>.

<u>P. 32</u>, <u>lines 5-9 from below</u>. Cf. the preceding correction. In the general case the proof in the text should be modified as follows. Replace $\frac{1}{2}\mathcal{U}_0$ by $\mathcal{U}_0'$, where

$$\mathcal{U}_0' + \mathcal{U}_0' \subset \mathcal{U}_0.$$

Then replace $(\frac{1}{2})^2\mathcal{U}_0$ by $\mathcal{U}_0''$, where $\mathcal{U}_0''$ is so chosen that

$$\mathcal{U}_0'' + \mathcal{U}_0'' \subset \mathcal{U}_0'$$

and so on, by induction: $\mathcal{U}_0^{(n)}$ having been chosen, take $\mathcal{U}_0^{(n+1)}$ such that

$$\mathcal{U}_0^{(n+1)} + \mathcal{U}_0^{(n+1)} \subset \mathcal{U}_0^{(n)}$$

as replacement for $(\frac{1}{2})^{n+1}\mathcal{U}_0$ in the text. Then we have

$$\mathcal{U}_0' + \mathcal{U}_0'' + \dots + \mathcal{U}_0^{(n)} \subset \mathcal{U}_0 \qquad (n = 1,2,\dots);$$

in fact, even

$$\Sigma_{1\leq j\leq n}\, \mathcal{U}_0^{(j)} + \mathcal{U}_0^{(n)} \subset \mathcal{U}_0,$$

by induction. With this modification, the proof in the text applies to the general case.

P. 36, last line of the text. REPLACE: smooth manifold BY: smooth (n-1)-dimensional manifold

P. 40, line 10. REPLACE: $\nu = 2$ BY: $\nu = 1$

P. 40, line 12. REPLACE: $\mathcal{F}^1_{\frac{1}{2}}(\mathbb{R}^2)$ BY: $\mathcal{F}^1_{\frac{1}{2}}(\mathbb{R})$

P. 58, line 6 from below. REPLACE: $\leqslant$ BY: $=$

P. 61, line 7 from below. REPLACE: The function BY: The continuous function

P. 70, line 4 from below. REPLACE: $f(\dot{x})$ BY: $\dot{f}(\dot{x})$

P. 74, line 11. REPLACE: footnote), p. 76. BY: footnote, p. 76).

P. 77, line 2. REPLACE: $[\mu^* \star f]^*$ BY: $[\mu^* \star f^*]^*$

P. 81, lines 6-7 from below. The proof may also be concluded in the following way. Since $g \in L^1(G)$ is arbitrary, we may take g as approximate left unit for f, whence $\langle f, \phi \rangle = 0$. This method uses only the formula (6) in place of (7).

P. 91, line 4 from below. REPLACE: $(\lambda_n)_{>0}$ BY: $(\lambda_n)_{n\geqslant 0}$

P. 95, line 13. REPLACE: is a topological BY: is also topological

P. 120, line 9. REPLACE: [108] BY: [108, Theorem 2 and Corollary 2]

P. 120, line 14. An inductive construction has since been given: see E. Hewitt and K.A. Ross, Abstract harmonic analysis II (Springer-Verlag, 1970), pp. 429-431.

P. 122, line 3 from below. REPLACE: this is shown as in 4.4 (i) BY: compare 4.4 (i) and 5.4

P. 129, line 17. DELETE: , but this is contained in 2.4 below

P. 130, line 10 from below. REPLACE: , and BY: holds and

P. 131, § 2.8. A simple example of a Segal algebra which does not have the property in question has been given by J. Cigler, Nederl. Akad. Wetensch. Indag. Math. 31 (1969), 273-282, especially pp. 276-277.

P. 146, line 6. The words 'of bounded L^1_w-norm' are extraneous to the argument and may be deleted.

P. 146, line 18. REPLACE: [111, I, Theorem 3] BY: [111, I, Theorem 4]

P. 160, line 5 from below. Note that the mapping of G onto G_1 defined by $x \to g$ $(x = gh)$ is continuous by hypothesis, and it is readily seen to be open.

P. 160, line 12 from below. REPLACE: the first footnote BY: the references in the first footnote

P. 167, line 16. REPLACE: Chapter III BY: Chapter 3

P. 171, line 20. REPLACE: $|f_n| \leq g$ BY $|f_n(x)| \leq |g(x)|$ a.e.

P. 171, line 12 from below. REPLACE: Now $g = \Sigma_{n\geq 1}|g_n|$ is in $L^p(G)$ and $|g_n| \leq g$, $n \geq 1$ BY: Now $\Sigma_{n\geq 1}|g_n|$ converges in $L^p(G)$, hence $|g_n(x)| \leq$ $\leq g(x)$ a.e. $(n \geq 1)$ for suitable $g \in L^p(G)$

P. 176, line 6 from below. In the sum at the end of the line the subscript y' should read y'_j.

P. 191, Reference 42. REPLACE: 461-2 BY: 431-2

P. 192, Reference 58. REPLACE: University of São Paulo BY: Sociedade de Matemática, São Paulo

P. 195, ADDITIONAL PUBLICATIONS:

Emerson, W.R. and Greenleaf, F.P. Covering properties and Følner conditions for locally compact groups. Math. Z. 102, 370-84 (1967).

Gilbert, J.E. Convolution operators on $L^p(G)$ and properties of locally compact groups. Pacific J. Math. 24, 257-68 (1968).

Leptin, H. Sur l'algèbre de Fourier d'une groupe localement compact. C.R. Acad. Sci. Paris 266, 1180-2 (1968).

BIBLIOGRAPHY

BURNHAM, J.T.: [1] Notes on subalgebras of Banach algebras with applications to harmonic analysis (mimeographed), 1969.

– [2] Notes on Banach algebras, 2nd edition (mimeographed), 1970.

CIGLER, J.: [1] Normed ideals in $L^1(G)$. Nederl. Akad. Wetensch. Indag. Math. 31 (1969), 273-282.

GILBERT, J.E.: [1] On projections of $L^\infty(G)$ onto translation-invariant subspaces. Proc. London Math. Soc. 19 (1969), 69-88.

ITÔ, T. and I. AMEMIYA: [1] A simple proof of the theorem of P.J. Cohen. Bull. Amer. Math. Soc. 70 (1964), 774-776.

LIU, T., A van ROOIJ and J. WANG: [1] Projections and approximate identities for ideals in group algebras (Preprint, 1970).

– [2] Bounded approximate identities in ideals of commutative group algebras (Preprint, 1970).

MALLIAVIN, P.: [1] Impossibilité de la synthèse spectrale sur les groupes abéliens non compacts. Inst. Hautes Etudes Publ. Math. No. 2 (1959), 61-68.

REITER, H.: [1] Contributions to harmonic analysis, IV. Math. Ann. 135 (1958), 467-476.

– [2] Sur certains idéaux dans $L^1(G)$. C.R. Acad. Sci. Paris 267 (1968), 882-885.

ROSENTHAL, H.P.: [1] On the existence of approximate identities in ideals of group algebras. Ark. Mat. 7 (1967), 185-191.

SCHREIBER, B.M.: [1] On the coset ring and strong Ditkin sets. Pacific J. Math. 32 (1970), 805-812.

WEIL, A.: [1] L'intégration dans les groupes topologiques et ses applications, 2nd edition (Hermann, Paris, 1953).

Lecture Notes in Mathematics

Comprehensive leaflet on request

Vol. 38: R. Berger, R. Kiehl, E. Kunz und H.-J. Nastold, Differentialrechnung in der analytischen Geometrie IV, 134 Seiten. 1967 DM 12,–

Vol. 39: Séminaire de Probabilités I. II, 189 pages. 1967. DM 14,–

Vol. 40: J. Tits, Tabellen zu den einfachen Lie Gruppen und ihren Darstellungen. VI, 53 Seiten. 1967. DM 6.80

Vol. 41: A. Grothendieck, Local Cohomology. VI, 106 pages. 1967. DM 10,–

Vol. 42: J. F. Berglund and K. H. Hofmann, Compact Semitopological Semigroups and Weakly Almost Periodic Functions. VI, 160 pages. 1967. DM 12,–

Vol. 43: D. G. Quillen, Homotopical Algebra. VI, 157 pages. 1967. DM 14,–

Vol. 44: K. Urbanik, Lectures on Prediction Theory. IV, 50 pages. 1967. DM 5,80

Vol. 45: A. Wilansky, Topics in Functional Analysis. VI, 102 pages. 1967. DM 9,60

Vol. 46: P. E. Conner, Seminar on Periodic Maps. IV, 116 pages. 1967. DM 10,60

Vol. 47: Reports of the Midwest Category Seminar I. IV, 181 pages. 1967. DM 14,80

Vol. 48: G. de Rham, S. Maumary et M. A. Kervaire, Torsion et Type Simple d'Homotopie. IV, 101 pages. 1967. DM 9,60

Vol. 49: C. Faith, Lectures on Injective Modules and Quotient Rings. XVI, 140 pages. 1967. DM 12,80

Vol. 50: L. Zalcman, Analytic Capacity and Rational Approximation. VI, 155 pages. 1968. DM 13.20

Vol. 51: Séminaire de Probabilités II. IV, 199 pages. 1968. DM 14,–

Vol. 52: D. J. Simms, Lie Groups and Quantum Mechanics. IV, 90 pages. 1968. DM 8,–

Vol. 53: J. Cerf, Sur les difféomorphismes de la sphère de dimension trois (Γ_4= O). XII, 133 pages. 1968. DM 12,–

Vol. 54: G. Shimura, Automorphic Functions and Number Theory. VI, 69 pages. 1968. DM 8,–

Vol. 55: D. Gromoll, W. Klingenberg und W. Meyer, Riemannsche Geometrie im Großen. VI, 287 Seiten. 1968. DM 20,–

Vol. 56: K. Floret und J. Wloka, Einführung in die Theorie der lokalkonvexen Räume. VIII, 194 Seiten. 1968. DM 16,–

Vol. 57: F. Hirzebruch und K. H. Mayer, O (n)-Mannigfaltigkeiten, exotische Sphären und Singularitäten. IV, 132 Seiten. 1968. DM 10,80

Vol. 58: Kuramochi Boundaries of Riemann Surfaces. IV, 102 pages. 1968. DM 9,60

Vol. 59: K. Jänich, Differenzierbare G-Mannigfaltigkeiten. VI, 89 Seiten. 1968. DM 8,–

Vol. 60: Seminar on Differential Equations and Dynamical Systems. Edited by G. S. Jones. VI, 106 pages. 1968. DM 9,60

Vol. 61: Reports of the Midwest Category Seminar II. IV, 91 pages. 1968. DM 9,60

Vol. 62: Harish-Chandra, Automorphic Forms on Semisimple Lie Groups X, 138 pages. 1968. DM 14,–

Vol. 63: F. Albrecht, Topics in Control Theory. IV, 65 pages. 1968. DM 6,80

Vol. 64: H. Berens, Interpolationsmethoden zur Behandlung von Approximationsprozessen auf Banachräumen. VI, 90 Seiten. 1968. DM 8,–

Vol. 65: D. Kölzow, Differentiation von Maßen. XII, 102 Seiten. 1968. DM 8,–

Vol. 66: D. Ferus, Totale Absolutkrümmung in Differentialgeometrie und -topologie. VI, 85 Seiten. 1968. DM 8,–

Vol. 67: F. Kamber and P. Tondeur, Flat Manifolds. IV, 53 pages. 1968. DM 5,80

Vol. 68: N. Boboc et P. Mustată, Espaces harmoniques associés aux opérateurs différentiels linéaires du second ordre de type elliptique. VI, 95 pages. 1968. DM 8,60

Vol. 69: Seminar über Potentialtheorie. Herausgegeben von H. Bauer. VI, 180 Seiten. 1968. DM 14,80

Vol. 70: Proceedings of the Summer School in Logic. Edited by M. H. Löb. IV, 331 pages. 1968. DM 20,–

Vol. 71: Séminaire Pierre Lelong (Analyse), Année 1967 – 1968. VI, 190 pages. 1968. DM 14,

Vol. 72: The Syntax and Semantics of Infinitary Languages. Edited by J. Barwise. IV, 268 pages. 1968. DM 18,–

Vol. 73: P. E. Conner, Lectures on the Action of a Finite Group. IV, 123 pages. 1968. DM 10,–

Vol. 74: A. Fröhlich, Formal Groups. IV, 140 pages. 1968. DM 12,–

Vol. 75: G. Lumer, Algèbres de fonctions et espaces de Hardy. VI, 80 pages. 1968. DM 8,–

Vol. 76: R. G. Swan, Algebraic K-Theory. IV, 262 pages. 1968. DM 18,–

Vol. 77: P.-A. Meyer, Processus de Markov: la frontière de Martin. IV, 123 pages. 1968. DM 10,–

Vol. 78: H. Herrlich, Topologische Reflexionen und Coreflexionen. XVI, 166 Seiten. 1968. DM 12,–

Vol. 79: A. Grothendieck, Catégories Cofibrées Additives et Complexe Cotangent Relatif. IV, 167 pages. 1968. DM 12,–

Vol. 80: Seminar on Triples and Categorical Homology Theory. Edited by B. Eckmann. IV, 398 pages. 1969. DM 20,–

Vol. 81: J.-P. Eckmann et M. Guenin, Méthodes Algébriques en Mécanique Statistique. VI, 131 pages. 1969. DM 12,–

Vol. 82: J. Wloka, Grundräume und verallgemeinerte Funktionen. VIII, 131 Seiten. 1969. DM 12,–

Vol. 83: O. Zariski, An Introduction to the Theory of Algebraic Surfaces. IV, 100 pages. 1969. DM 8,–

Vol. 84: H. Lüneburg, Transitive Erweiterungen endlicher Permutationsgruppen. IV, 119 Seiten. 1969. DM 10.–

Vol. 85: P. Cartier et D. Foata, Problèmes combinatoires de commutation et réarrangements. IV, 88 pages. 1969. DM 8,–

Vol. 86: Category Theory, Homology Theory and their Applications I. Edited by P. Hilton. VI, 216 pages. 1969. DM 16,–

Vol. 87: M. Tierney, Categorical Constructions in Stable Homotopy Theory. IV, 65 pages. 1969. DM 6,–

Vol. 88: Séminaire de Probabilités III. IV, 229 pages. 1969. DM 18,–

Vol. 89: Probability and Information Theory. Edited by M. Behara, K. Krickeberg and J. Wolfowitz. IV, 256 pages. 1969. DM 18,–

Vol. 90: N. P. Bhatia and O. Hajek, Local Semi-Dynamical Systems. II, 157 pages. 1969. DM 14,–

Vol. 91: N. N. Janenko, Die Zwischenschrittmethode zur Lösung mehrdimensionaler Probleme der mathematischen Physik. VIII, 194 Seiten. 1969. DM 16,80

Vol. 92: Category Theory, Homology Theory and their Applications II. Edited by P. Hilton. V, 308 pages. 1969. DM 20,–

Vol. 93: K. R. Parthasarathy, Multipliers on Locally Compact Groups. III, 54 pages. 1969. DM 5,60

Vol. 94: M. Machover and J. Hirschfeld, Lectures on Non-Standard Analysis. VI, 79 pages. 1969. DM 6,–

Vol. 95: A. S. Troelstra, Principles of Intuitionism. II, 111 pages. 1969. DM 10,–

Vol. 96: H.-B. Brinkmann und D. Puppe, Abelsche und exakte Kategorien, Korrespondenzen. V, 141 Seiten. 1969. DM 10,–

Vol. 97: S. O. Chase and M. E. Sweedler, Hopf Algebras and Galois theory. II, 133 pages. 1969. DM 10,–

Vol. 98: M. Heins, Hardy Classes on Riemann Surfaces. III, 106 pages. 1969. DM 10,–

Vol. 99: Category Theory, Homology Theory and their Applications III. Edited by P. Hilton. IV, 489 pages. 1969. DM 24,–

Vol. 100: M. Artin and B. Mazur, Etale Homotopy. II, 196 Seiten. 1969. DM 12,–

Vol. 101: G. P. Szegö et G. Treccani, Semigruppi di Trasformazioni Multivoche. VI, 177 pages. 1969. DM 14,–

Vol. 102: F. Stummel, Rand- und Eigenwertaufgaben in Sobolewschen Räumen. VIII, 386 Seiten. 1969. DM 20,–

Vol. 103: Lectures in Modern Analysis and Applications I. Edited by C. T. Taam. VII, 162 pages. 1969. DM 12,–

Vol. 104: G. H. Pimbley, Jr., Eigenfunction Branches of Nonlinear Operators and their Bifurcations. II, 128 pages. 1969. DM 10,–

Vol. 105: R. Larsen, The Multiplier Problem. VII, 284 pages. 1969. DM 18,–

Vol. 106: Reports of the Midwest Category Seminar III. Edited by S. Mac Lane. III, 247 pages. 1969. DM 16,–

Vol. 107: A. Peyerimhoff, Lectures on Summability. III, 111 pages. 1969. DM 8,–

Vol. 108: Algebraic K-Theory and its Geometric Applications. Edited by R. M.F. Moss and C. B. Thomas. IV, 86 pages. 1969. DM 6,–

Vol. 109: Conference on the Numerical Solution of Differential Equations. Edited by J. Ll. Morris. VI, 275 pages. 1969. DM 18,–

Vol. 110: The Many Facets of Graph Theory. Edited by G. Chartrand and S. F. Kapoor. VIII, 290 pages. 1969. DM 18,–

Vol. 111: K. H. Mayer, Relationen zwischen charakteristischen Zahlen. III, 99 Seiten. 1969. DM 8,–

Vol. 112: Colloquium on Methods of Optimization. Edited by N. N. Moiseev. IV, 293 pages. 1970. DM 18,–

Vol. 113: R. Wille, Kongruenzklassengeometrien. III, 99 Seiten. 1970. DM 8,–

Vol. 114: H. Jacquet and R. P. Langlands, Automorphic Forms on GL (2). VII, 548 pages. 1970. DM 24,–

Vol. 115: K. H. Roggenkamp and V. Huber-Dyson, Lattices over Orders I. XIX, 290 pages. 1970. DM 18,–

Vol. 116: Sèminaire Pierre Lelong (Analyse) Année 1969. IV, 195 pages. 1970. DM 14,–

Vol. 117: Y. Meyer, Nombres de Pisot, Nombres de Salem et Analyse Harmonique. 63 pages. 1970. DM 6.–

Vol. 118: Proceedings of the 15th Scandinavian Congress, Oslo 1968. Edited by K. E. Aubert and W. Ljunggren. IV, 162 pages. 1970. DM 12,–

Vol. 119: M. Raynaud, Faisceaux amples sur les schémas en groupes et les espaces homogènes. III, 219 pages. 1970. DM 14,–

Vol. 120: D. Siefkes, Büchi's Monadic Second Order Successor Arithmetic. XII, 130 Seiten. 1970. DM 12,–

Vol. 121: H. S. Bear, Lectures on Gleason Parts. III, 47 pages. 1970. DM 6,–

Vol. 122: H. Zieschang, E. Vogt und H.-D. Coldewey, Flächen und ebene diskontinuierliche Gruppen. VIII, 203 Seiten. 1970. DM 16,–

Vol. 123: A. V. Jategaonkar, Left Principal Ideal Rings. VI, 145 pages. 1970. DM 12,–

Vol. 124: Séminare de Probabilités IV. Edited by P. A. Meyer. IV, 282 pages. 1970. DM 20,–

Vol. 125: Symposium on Automatic Demonstration. V, 310 pages. 1970. DM 20,–

Vol. 126: P. Schapira, Théorie des Hyperfonctions. XI, 157 pages. 1970. DM 14,–

Vol. 127: I. Stewart, Lie Algebras. IV, 97 pages. 1970. DM 10,–

Vol. 128: M. Takesaki, Tomita's Theory of Modular Hilbert Algebras and its Applications. II, 123 pages. 1970. DM 10,–

Vol. 129: K. H. Hofmann, The Duality of Compact Semigroups and C*- Bigebras. XII, 142 pages. 1970. DM 14,–

Vol. 130: F. Lorenz, Quadratische Formen über Körpern. II, 77 Seiten. 1970. DM 8,–

Vol. 131: A Borel et al., Seminar on Algebraic Groups and Related Finite Groups. VII, 321 pages. 1970. DM 22,–

Vol. 132: Symposium on Optimization. III, 348 pages. 1970. DM 22,–

Vol. 133: F. Topsøe, Topology and Measure. XIV, 79 pages. 1970. DM 8,–

Vol. 134: L. Smith, Lectures on the Eilenberg-Moore Spectral Sequence. VII, 142 pages. 1970. DM 14,–

Vol. 135: W. Stoll, Value Distribution of Holomorphic Maps into Compact Complex Manifolds. II, 267 pages. 1970. DM 18,–

Vol. 136: M. Karoubi et al., Séminaire Heidelberg-Saarbrücken-Strasbuorg sur la K-Théorie. IV, 264 pages. 1970. DM 18,–

Vol. 137: Reports of the Midwest Category Seminar IV. Edited by S. MacLane. III, 139 pages. 1970. DM 12,–

Vol. 138: D. Foata et M. Schützenberger, Théorie Géométrique des Polynômes Eulériens. V, 94 pages. 1970. DM 10,–

Vol. 139: A. Badrikian, Séminaire sur les Fonctions Aléatoires Linéaires et les Mesures Cylindriques. VII, 221 pages. 1970. DM 18,–

Vol. 140: Lectures in Modern Analysis and Applications II. Edited by C. T. Taam. VI, 119 pages. 1970. DM 10,–

Vol. 141: G. Jameson, Ordered Linear Spaces. XV, 194 pages. 1970. DM 16,–

Vol. 142: K. W. Roggenkamp, Lattices over Orders II. V, 388 pages. 1970. DM 22,–

Vol. 143: K. W. Gruenberg, Cohomological Topics in Group Theory. XIV, 275 pages. 1970. DM 20,–

Vol. 144: Seminar on Differential Equations and Dynamical Systems, II. Edited by J. A. Yorke. VIII, 268 pages. 1970. DM 20,–

Vol. 145: E. J. Dubuc, Kan Extensions in Enriched Category Theory. XVI, 173 pages. 1970. DM 16,–

Vol. 146: A. B. Altman and S. Kleiman, Introduction to Grothendieck Duality Theory. II, 192 pages. 1970. DM 18,–

Vol. 147: D. E. Dobbs, Cech Cohomological Dimensions for Commutative Rings. VI, 176 pages. 1970. DM 16,–

Vol. 148: R. Azencott, Espaces de Poisson des Groupes Localement Compacts. IX, 141 pages. 1970. DM 14,–

Vol. 149: R. G. Swan and E. G. Evans, K-Theory of Finite Groups and Orders. IV, 237 pages. 1970. DM 20,–

Vol. 150: Heyer, Dualität lokalkompakter Gruppen. XIII, 372 Seiten. 1970. DM 20,–

Vol. 151: M. Demazure et A. Grothendieck, Schémas en Groupes I. (SGA 3). XV, 562 pages. 1970. DM 24,–

Vol. 152: M. Demazure et A. Grothendieck, Schémas en Groupes II. (SGA 3). IX, 654 pages. 1970. DM 24,–

Vol. 153: M. Demazure et A. Grothendieck, Schémas en Groupes III. (SGA 3). VIII, 529 pages. 1970. DM 24,–

Vol. 154: A. Lascoux et M. Berger, Variétés Kähleriennes Compactes. VII, 83 pages. 1970. DM 8,–

Vol. 155: Several Complex Variables I, Maryland 1970. Edited by J. Horváth. IV, 214 pages. 1970. DM 18,–

Vol. 156: R. Hartshorne, Ample Subvarieties of Algebraic Varieties. XIV, 256 pages. 1970. DM 20,–

Vol. 157: T. tom Dieck, K. H. Kamps und D. Puppe, Homotopietheorie. VI, 265 Seiten. 1970. DM 20,–

Vol. 158: T. G. Ostrom, Finite Translation Planes. IV. 112 pages. 1970. DM 10,–

Vol. 159: R. Ansorge und R. Hass. Konvergenz von Differenzenverfahren für lineare und nichtlineare Anfangswertaufgaben. VIII, 145 Seiten. 1970. DM 14,–

Vol. 160: L. Sucheston, Constributions to Ergodic Theory and Probability. VII, 277 pages. 1970. DM 20,–

Vol. 161: J. Stasheff, H-Spaces from a Homotopy Point of View. VI, 95 pages. 1970. DM 10,–

Vol. 162: Harish-Chandra and van Dijk, Harmonic Analysis on Reductive p-adic Groups. IV, 125 pages. 1970. DM 12,–

Vol. 163: P. Deligne, Equations Différentielles à Points Singuliers Reguliers. III, 133 pages. 1970. DM 12,–

Vol. 164: J. P. Ferrier, Seminaire sur les Algebres Complètes. II, 69 pages. 1970. DM 8,–

Vol. 165: J. M. Cohen, Stable Homotopy. V, 194 pages. 1970. DM 16,–

Vol. 166: A. J. Silberger, PGL_2 over the p-adics: its Representations, Spherical Functions, and Fourier Analysis. VII, 202 pages. 1970. DM 18,–

Vol. 167: Lavrentiev, Romanov and Vasiliev, Multidimensional Inverse Problems for Differential Equations. V, 59 pages. 1970. DM 10,–

Vol. 168: F. P. Peterson, The Steenrod Algebra and its Applications: A conference to Celebrate N. E. Steenrod's Sixtieth Birthday. VII, 317 pages. 1970. DM 22,–

Vol. 169: M. Raynaud, Anneaux Locaux Henséliens. V, 129 pages. 1970. DM 12,–

Vol. 170: Lectures in Modern Analysis and Applications III. Edited by C. T. Taam. VI, 213 pages. 1970. DM 18,–

Vol. 171: Set-Valued Mappings, Selections and Topological Properties of 2^X. Edited by W. M. Fleischman. X, 110 pages. 1970. DM 12,–

Vol. 172: Y.-T. Siu and G. Trautmann, Gap-Sheaves and Extension of Coherent Analytic Subsheaves. V, 172 pages. 1971. DM 16,–

Vol. 173: J. N. Mordeson and B. Vinograde, Structure of Arbitrary Purely Inseparable Extension Fields. IV, 138 pages. 1970. DM 14,–

Vol. 174: B. Iversen, Linear Determinants with Applications to the Picard Scheme of a Family of Algebraic Curves. VI, 69 pages. 1970. DM 8,–

Vol. 175: M. Brelot, On Topologies and Boundaries in Potential Theory. VI, 176 pages. 1971. DM 18,–

Vol. 176: H. Popp, Fundamentalgruppen algebraischer Mannigfaltigkeiten. IV, 154 Seiten. 1970. DM 16,–

Vol. 177: J. Lambek, Torsion Theories, Additive Semantics and Rings of Quotients. VI, 94 pages. 1971. DM 12,–

Vol. 178: Th. Bröcker und T. tom Dieck, Kobordismentheorie. XVI, 191 Seiten. 1970. DM 18,–

Vol. 179: Seminaire Bourbaki – vol. 1968/69. Exposés 347-363. IV. 295 pages. 1971. DM 22,–

Vol. 180: Séminaire Bourbaki – vol. 1969/70. Exposés 364-381. IV, 310 pages. 1971. DM 22,–

Vol. 181: F. DeMeyer and E. Ingraham, Separable Algebras over Commutative Rings. V, 157 pages. 1971. DM 16.–